Carolina Machado, J. Paulo Davim (Eds.)
Productivity and Organizational Management

Also of interest

Innovation Management.
In Research and Industry, 2. ED
Carolina Machado and J. Paulo Davim, 2025
ISBN 978-3-11-914707-1, e-ISBN (PDF) 978-3-11-220669-0

Product and Process Design.
Driving Sustainable Innovation
Jan Harmsen, André B. de Haan and Pieter L. J. Swinkels, 2024
ISBN 978-3-11-078206-6, e-ISBN (PDF) 978-3-11-078212-7

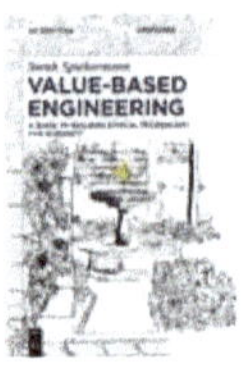

Value-Based Engineering.
A Guide to Building Ethical Technology for Humanity
Sarah Spiekermann, 2023
ISBN 978-3-11-079336-9, e-ISBN (PDF) 978-3-11-079338-3

Intelligent Educational Robots.
Toward Personalized Learning Environments
Edited by: Stamatios Papadakis and Georgios Lampropoulos, 2024
ISBN 978-3-11-135206-0, e-ISBN (PDF) 978-3-11-135269-5

Engineering Risk Management
Thierry Meyer and Genserik Reniers, 2022
ISBN 978-3-11-066531-4, e-ISBN (PDF) 978-3-11-066533-8

Productivity and Organizational Management

Management Tools, Human Resource Management, Contemporary Engineers

2nd, Revised and Extended Edition

Edited by Carolina Machado and J. Paulo Davim

DE GRUYTER

Editors
Prof. Carolina Machado
School of Economics, Management and Political Science
University of Minho
Campus Gualtar
4710-057 Braga
Portugal
carolina@eeg.uminho.pt

Prof. Dr. J. Paulo Davim
Department of Mechanical Engineering
University of Aveiro
Campus Santiago
3810-193 Aveiro
Portugal
pdavim@ua.pt

ISBN 978-3-11-914732-3
e-ISBN (PDF) 978-3-11-220677-5
e-ISBN (EPUB) 978-3-11-220700-0

Library of Congress Control Number: 2025941265

Bibliographic information published by the Deutsche Nationalbibliothek
The Deutsche Nationalbibliothek lists this publication in the Deutsche Nationalbibliografie; detailed bibliographic data are available on the Internet at http://dnb.dnb.de.

Cover image: Creative-Touch/DigitalVision Vectors/Getty Images
Typesetting: Integra Software Services Pvt. Ltd.

www.degruyterbrill.com
Questions about General Product Safety Regulation:
productsafety@degruyterbrill.com

Opening note

Titled its first edition *Productivity and Organizational Management*, and due to the positive answer that the market has been showing, the editors decided to update and expand this book in a second edition now with the new title *Productivity and Organizational Management – Management Tools, Human Resource Management, Contemporary Engineers*. The changes introduced involved a brief update of the text and the introduction of ALT TEXT throughout the different chapters. This introduction was made due to the new legal requirement (European Accessibility Act). All e-book publications must be accessible to everyone, which implements the accessibility of graphic content (namely images) with the help of alternative texts describing these images.

Being conscious of the evolution that the topic under study has been assuming over time, it was our understanding to accompany these updates with the introduction of two new chapters where the issues of productivity and organizational management are reinforced. The aim is to help professionals and all kinds of readers interested in acquiring and strengthening their knowledge in this field, so that they can, with this new, updated, and expanded edition, acquire the necessary knowledge to respond more efficiently and effectively and, more than that, anticipate the challenges that organizations are facing with regard to their performance and productivity. We thus come across the insertion of Chapter 7 (with the inclusion of this new chapter, chapter 7 of the first edition has been moved to Chapter 8), entitled "Competencies management in an SME: a lever for greater productivity," and Chapter 9, where the book closes with a final note on "Productivity and management: a bibliometric analysis."

The editors would like to thank De Gruyter and all the authors and coauthors for the contributions and improvements implemented in the chapters, as well as all the readers, students, and professionals in the area of management and engineering, not only for the good acceptance that the book has had but also for their contribution to the improvement and enhancement of the work.

The Editors
Carolina Machado
University of Minho

J. Paulo Davim
University of Aveiro

July 2025

https://doi.org/10.1515/9783112206775-202

Preface

Integrated in an environment characterized by strong levels of competitiveness and deep challenges, today's organizations are facing a continuous and highly complex process of change. Conscious of this reality, this book, entitled *Productivity and Organizational Management – Management Tools, Human Resource Management, Contemporary Engineers*, covers the issues related to the new advances in productivity and organizational management that result from these environment progressive requests. As a consequence of social, technological, political, and economic changes, organizations need to be more and more proactive in order to anticipate these changes. Indeed, the field of work and organizational management becomes highly complex, claiming management for more effective and proactive strategies. Taking into account this reality, this book looks to provide support to academics and researchers, as well as those operating in the management field who need to deal with policies and strategies related to work issues. Presently, we live in a time in which there is a vast amount of uncertainty and ambiguity about the direction in which organizations are moving. While many advances have been made in understanding productivity, the social and organizational context remains problematic. Managers, engineers, and other professionals still ask questions about the best strategies to obtain high productivity levels. It remains a mystery to know what is the best work organization and design, compatible with the desirable productivity levels. Interdisciplinary perspectives to further our knowledge and understanding of productivity and related change processes and work practices are needed. They will contribute to a better merger and interrelationship among organization, management, and employee needs in order to increase efficiency, productivity, and profitability. Effective work practices and good employee relations are a real necessity of today's organizations, as they can help reduce absenteeism, turnover, and organizational costs, conducting to high levels of commitment, effectiveness, performance, as well as productivity. Addressing these questions, this book focuses on the implications of those changes in productivity and organizational management. It explores the models, tools, and processes used by organizations in order to help managers and all managerial professionals become better prepared to face the challenges and changes in work and, consequently, in the way how to manage today's organizations in order to obtain high levels of productivity.

Looking to share knowledge, through debate and information exchange, about productivity and organizational management, this book is divided into nine chapters. Chapter 1 covers "the measurement of human capital and its relation to the generation of business value: empirical study." Chapter 2 discusses "a case of certified units in a Portuguese University: interactions of ISO 9000 norms with HRM practices, employee performance, and employee satisfaction." Chapter 3 contains information on "management tools for supporting productivity in organizations: empirical evidence from Slovenia." Chapter 4 describes "economic and social efficiency: the case for inverting the principle of productivity in public services." Subsequently, Chapter 5 cov-

https://doi.org/10.1515/9783112206775-203

ers "human resource management in the health System: in the never-ending quest for productivity improvement – the Portuguese National Health Service case." Chapter 6 describes the "critical role of managerial competencies in productivity enhancement interventions: an HRM perspective." Finally, in Chapter 7, "competencies management in an SME: a lever for greater productivity" are analyzed, just followed by Chapter 8 in which "prospective characteristics of contemporary engineer (by the approach of mechanical engineering): contribution and role of the mechanical engineer to the organization management and productivity" is presented. The book closes with Chapter 9, where we can find relevant information through "productivity and management: a bibliometric analysis."

Considered an excellent opportunity to participate in an exchange of information, ideas, and opinions about productivity and organizational management, we can say that this book is designed to increase the knowledge and understanding of all those – HR managers, managers, engineers, entrepreneurs, strategists, practitioners, academics, or researchers – involved with productivity issues, in all kinds of organizations and activity sectors.

The editors acknowledge their gratitude to De Gruyter for this opportunity and for their professional support. Finally, we would like to thank to all chapter authors for their interest and availability to work on this project.

Carolina Machado
Braga, Portugal

J. Paulo Davim
Aveiro, Portugal

Contents

Opening note —— V

Preface —— VII

About the editors —— XI

List of authors —— XIII

Santiago Gutiérrez-Broncano, Mercedes Rubio-Andrés
and Juan Carlos Zapata Valencia
1 The measurement of human capital and its relation to the generation of business value: empirical study —— 1

Ana Paula Ferreira and Marta Lopes
2 A case of certified units in a Portuguese University: interactions of ISO 9000 norms with HRM practices, employee performance, and employee satisfaction —— 21

Zlatko Nedelko and Vojko Potocan
3 Management tools for supporting productivity in organizations: empirical evidence from Slovenia —— 49

Teresa Carla Oliveira and Stuart Holland
4 Economic and social efficiency: the case for inverting the principle of productivity in public services —— 77

Diana Santos Fernandes and Carolina Feliciana Machado
5 Human resource management in the health system: in the never-ending quest for productivity improvement —— 109

Alamuri Suryanarayana
6 Critical role of managerial competencies in productivity enhancement interventions: an HRM perspective —— 133

Sara Maria Mendes FERNANDES and Carolina Feliciana MACHADO
7 Competencies management in an SME: a lever for greater productivity —— 151

Emin Taner Elmas
8 Prospective characteristics of contemporary engineers (using the approach of mechanical engineering) — 167

J. Paulo Davim
9 Productivity and management: a bibliometric analysis — 183

Index — 187

About the editors

Carolina Machado received her PhD in management sciences (organizational and policies management area/human resource management) from the University of Minho in 1999, her master's degree in management (strategic human resource management) from the Technical University of Lisbon in 1994, and degree in business administration from the University of Minho in 1989. She is teaching human resource management subjects since 1989 at the University of Minho, an associate professor since 2004 (with habilitation since 2022), with experience and research interest areas in the fields of human resource management, international human resource management, human resource management in SMEs, training and development, emotional intelligence, management change, knowledge management, management/HRM in the digital age, Industry 5.0, sustainability, and higher education sustainability. She is head of the Human Resources Management Work Group at the School of Economics, Management and Political Science at the University of Minho, coordinator of Advanced Training Courses at the Interdisciplinary Centre of Social Sciences, member of the Interdisciplinary Centre of Social Sciences (CICS.NOVA.UMinho), University of Minho, as well as chief editor of the *International Journal of Applied Management Sciences and Engineering* (IJAMSE), guest editor of journals, book editor, and book series editor, as well as reviewer in different international prestigious journals. In addition, she has also published both as editor/coeditor and as author/coauthor of several books, book chapters, and articles in journals and conferences.
ORCID: http://orcid.org/0000-0002-9685-1576
E-mail: carolina@eeg.uminho.pt

J. Paulo Davim is a full professor at the University of Aveiro, Portugal. He is also distinguished as honorary professor in several universities/colleges/institutes in China, India, and Spain. He received his PhD in mechanical engineering, MSc in mechanical engineering (materials and manufacturing processes) and mechanical engineering degree (5 years) from the University of Porto (FEUP), the aggregate title (full habilitation) from the University of Coimbra, and the DSc (higher doctorate) from London Metropolitan University. He is fellow (FIET) of IET-London, fellow (FIMF) of IMF-Birmingham, and Eur Ing by Engineers Europe FEANI-Brussels. He is also senior specialist engineer by the Portuguese Order of Engineers with an MBA and specialist titles in industrial management as well as in metrology. He has more than 35 years of teaching and research experience in manufacturing, materials, mechanical and industrial engineering, with special emphasis in machining, tribology, and surface engineering. He has also interest in design, management, sustainability, Industry 5.0, engineering education, and higher education for sustainability. He has guided large numbers of postdoc, PhD, and master's students, as well as coordinated and participated in several financed research projects. He has received several scientific awards and honors. He has worked as evaluator of projects for ERC-European Research Council and other international research agencies, as well as examiner of PhD thesis for many universities in different countries. He is the editor in chief of several international journals, book series editor, and scientific advisory for many conferences. He has been listed in the World's Top 2% of Scientists by Stanford University study.
ORCID: https://orcid.org/0000-0002-5659-3111
E-mail: pdavim@ua.pt

https://doi.org/10.1515/9783112206775-205

List of authors

Santiago Gutiérrez-Broncano
University of Castilla-La Mancha
Castilla-La Mancha
Spain
Santiago.gutierrez@uclm.es
Chapter 1

Mercedes Rubio-Andrés
San Pablo CEU University
Madrid
Spain
Chapter 1

Juan Carlos Zapata Valencia
Pontificia Bolivariana University
Antioquia, Colombia
Chapter 1

Ana Paula Ferreira
Department of Management
School of Economics, Management and Political Science
University of Minho
Campus Gualtar, 4710-057 Braga
Portugal
aferreira@eeg.uminho.pt
Chapter 2

Marta Lopes
Department of Management
School of Economics, Management and Political Science
University of Minho
Campus Gualtar, 4710-057 Braga
Portugal
Chapter 2

Zlatko Nedelko
Faculty of Economics and Business
University of Maribor
Maribor
Slovenia
zlatko.nedelko@uni-mb.si
Chapter 3

Vojko Potocan
Faculty of Economics and Business
University of Maribor
Maribor
Slovenia
vojko.potocan@uni-mb.si
Chapter 3

Stuart Holland
Faculty of Economics
University of Coimbra
CEISUC – Centro de Estudos e Investigação em Saúde, University of Coimbra
(Institute for Social and European Studies, Hungary)
Av. Dias da Silva 165
3004-512 Coimbra
Portugal
Chapter 4

Teresa Carla Oliveira
Faculty of Economics
University of Coimbra
CEISUC – Centro de Estudos e Investigação em Saúde, University of Coimbra
Av. Dias da Silva 165
3004-512 Coimbra
Portugal
tcarla@fe.uc.pt
Chapter 4

Diana Santos Fernandes
Department of Management
School of Economics, Management and Political Science
University of Minho
Campus Gualtar, 4710-057 Braga
Portugal
Chapter 5

https://doi.org/10.1515/9783112206775-206

Carolina Feliciana Machado
Department of Management
School of Economics, Management and Political Science
University of Minho
Campus Gualtar, 4710-057 Braga
Portugal
carolina@eeg.uminho.pt
Chapters 5 and 7

Sara Maria Mendes Fernandes
Department of Management
School of Economics, Management and Political Science
University of Minho
Campus Gualtar, 4710-057 Braga
Portugal
Chapter 7

Alamuri Suryanarayana
Osmania University
Hyderabad, Telangana
India
professorsuryanarayana@gmail.com
Chapter 6

Emin Taner Elmas
Sistemas Technology and System Production Co.
Mansuroglu Mahallesi 263 Sok., Ozyegin Apt.
No. 10 K:3 D:7 35535
Bayrakli, Izmir
Turkey
emintanerelmas@gmail.com;
muhendis123@hotmail.com
Chapter 8

J. Paulo Davim
Department of Mechanical Engineering
University of Aveiro
Campus Santiago
3810-193 Aveiro
Portugal
pdavim@ua.pt
Chapter 9

Santiago Gutiérrez-Broncano, Mercedes Rubio-Andrés
and Juan Carlos Zapata Valencia

1 The measurement of human capital and its relation to the generation of business value: empirical study

Abstract: Although various empirical studies have attempted to measure the strategic contribution of human capital to a company, only a few use economic and financial indicators to measure that contribution. In this chapter, we attempt to demonstrate how a measurement variable of human capital investment has a significant impact on the generation of business value. We also examine the role of a given sector in this relationship. Using a sample of 23 companies belonging to the Asociación Colombiana de Relaciones de Trabajo (Colombian Association of Work Relations) (ASCORT), we find that the larger the investment in human capital, the greater the economic development of a company, especially those in the service sector. These findings open the door to new research linking the strategic development of companies to investments in human capital.

1.1 Introduction

Although most managers have the intuition that investment in employees typically results in additional value for their enterprise, it is unclear just how much additional value is generated [1].

"Our employees are the most important asset of the company" are words heard in most organizations. Despite this, few companies are able to prove that this asset indeed has an impact on the creation of added value for the company. In general, more emphasis is given to practices, such as employee satisfaction or investment in training; however, minimal importance is placed on the value generated by human capital.

Companies certainly wish to recover their investments in their employees and assume that employees will contribute beyond their own salaries and benefits, but it is difficult to determine just what their contribution consists of.

What might be the reaction of Wall Street analysts if a company were to announce that it was going to close one of its 10 production plants to reduce costs? If we assume a typical 10% cut in the workforce, including generous compensation and early retirement, would this be more attractive to shareholders? This is the great dilemma facing most companies today. Intangible assets, as well as everything related

https://doi.org/10.1515/9783112206775-001

to human capital, corporate social responsibility, reputation, and other items, are typically not factored into operating costs [2].

One might wonder whether investments in human resources reap financial rewards. The objective of this study is to identify variables in the measurement of human capital in enterprises that make it possible to establish a relationship between human capital and the generation of business value. To do this, we begin with a review of the literature on the importance of investing human capital and reveal the main indicators that allow us to measure the different organizational levels associated with human capital. This leads us to focus on those indicators that measure contributions at the strategic level. Finally, we carry out an analysis on 23 companies belonging to the Colombian Association of Work Relations (ASCORT) with the aim of studying the relationship between a company's investment in human capital and the value generated by such an investment.

1.2 Theoretical background and hypothesis

1.2.1 Human capital and its relationship to value creation

Human capital has been defined at the individual level as the combination of four factors: heredity, training, experience, and attitude [3]. It is the sum of aspects inherent in workers: skills, knowledge, experiences, habits, motivation, and even the energy that workers demonstrate in carrying out their duties, creating a specific culture in each company [4]. Under these conditions, employees are able to learn, innovate, stimulate, and make changes, as well as think creatively, all of which factors are vital for the future success of a company.

The human resource management literature is based on the premise that human resources are vital to an organization's strategy since, through their behavior, individuals have the potential to lay the foundation for strategy formulation and implementation [5]. In this way, they recognize that individuals' knowledge, skills, and abilities are insufficient to create value for an organization, unless they are used through individual behavior [6, 7].

The human capital of companies is not precisely the workers, but rather the workers are the owners of their own human capital, and they exhibit said behavior in different areas of their private lives, including their interactions with their families, their communities, their hobbies, their sports, and in their work [1]. In this paper, we accept that a firm does not possess human capital; rather, individuals do. Companies need a higher level of knowledge, skills, and abilities; however, a superior alignment between individual and organizational goals is crucial [8].

Furthermore, many papers and reports relate the management of human resources to a greater generation of business value. The literature on strategic management of human resources has found sufficient evidence supporting the idea that a high commitment practices in human resource management lead to greater organizational performance (e.g., [9–14]). This set of practices, which is aimed at the management and development of human capital in organizations, serves to establish long-term relationships between companies and their employees, thereby promoting the sharing of the same strategic goals as well as the creation of a strong commitment between both parties and their respective objectives. All of this contributes to better performance of the organization [15].

A survey on human capital carried out by IBM Business Consulting Services at more than a thousand companies from 47 different countries demonstrates that a well-designed and properly implemented human resource strategy can achieve a 35% increase in revenue per employee and a 12% decrease in absenteeism [16]. Conversely, those companies that reduced their workforce by more than 15% saw the price of their shares fall below the average level for their industry [17].

1.2.2 Levels of human capital measurement: the value chain of human talent

Both the literature on human resources management and enterprise managers themselves have always recognized the need to establish human capital measurement systems in companies [1, 18].

Traditionally, researchers have measured the quality of human resources management systems using data related to various personnel policies implemented in a company gathered from a single informant [19]. Thus, the impact of human resource practices in business outcomes has been significantly different from what such works suggest and may be even greater than indicated in the literature.

One detected error is located in what is known as the *percept–percept inflation*, which occurs when the information of both dependent and independent variables originate from the same source [20]. This type of error poses the least threat, thanks to the fact that most works obtained information from different sources [19].

Perhaps the more serious issues include those raised by Rush et al. [21], who established that respondents, either owing to faulty memory or because the requested data are unknown to them in their entirety, react by resorting to so-called implicit theories, which act as counsel when gaps in information are filled in a search for consistency among all data that have been produced. Thus, when information is requested by an informant, whether objective or subjective, it would seem logical to think that the respondent will react to information concerning the implementation of human resource policies in a way that is aligned with the same direction as the business results perceived by that informant.

Subsequent studies have supported this theory, demonstrating the need for some caution when it comes to measuring systems and human resource practices in enterprises [19]. For this reason, this work seeks alternative ways to measure the impact of human resource management systems in a company in a more objective way, as well as the impact those systems have on the generation of added value.

Historically, a variety of methods have been used to measure the productivity of the human factor. However, these methods have been mainly oriented toward efficiency in carrying out certain administrative tasks or related to the cost of productive activities [22]. The main problem that we face today is that the type of work carried out by the majority of employees and that generates greater business value is very different from traditional jobs; therefore, new formulas for the measurement, subsequent management, and improvement of value are required [23]. As shown in the latest report on global trends in human capital 2016 performed with more than 7,000 business leaders and human resource experts in 100 countries, the great challenges of global impact involve redesigning the workplace, the workforce, and work itself [24].

There is a need for systems that can be used to evaluate the human capital contribution to business value creation and that are focused on more than just the technical efficiency of finished tasks or operations. Bontis and Fitz-Enz [25] suggest that the measurement of human capital is critical and vital to the future success of companies. The environment in which a company is currently functioning and that is becoming increasingly more competitive requires all areas to address the growing demand for better results [26]. Most company CEOs ponder the impact of their training and development, reconciliation of work and family life, improvement of the quality of life at work, work flexibility programs, and other aspects of work. In summary, it is crucial to be aware of the economic impact of investments in human capital development carried out by enterprises as part of their overall productivity and whether this will lead to dramatic improvements in business strategy and generate additional profits [27].

We must be aware that no department is allowed to make an investment without prior analysis of future profitability. However, in the field of human resources, despite the fact that most managers recognize that said resource is the main asset of the company, it is difficult to predict what profitability a given investment will generate owing to the difficulty involved in its measurement.

Therefore, this issue adds to current accounting problems, since the majority of intangible assets go unnoticed with respect to accounting information because they do not appear on balance sheets. As a result, the majority of investments carried out for staffs are directly treated as costs. Thus, initially, a company will try to reduce costs, and the first programs to be cut will be those related to human factors since a clear and precise assessment of their impact on business results is unobtainable [1].

At this point, we may hesitate in asserting that human capital is an asset, but the definition given by the Financial Accounting Standards Board (FASB) [28] leaves no doubt that our theory is correct. The FASB establishes three criteria for such a definition it must be an economic resource that is able to generate a cash influx or reduce outflows; the organization should aim to have rights or privileges in connection with it; and that resource as well as the rights of the employees must be listed in the financial statements of the company. In summary, an asset is a resource that is controlled by the company and is used to produce additional future inflows or reduce outflows. As for employees, we cannot regard them as a company asset as such, however, in the case of human capital companies usually establish a series of agreements by which employees put their capacities and abilities at the disposal of companies to generate income flows [1]. Thus, their capacities and abilities become a company asset.

One might wonder whether we can really measure the human capital of a company and analyze employee contributions. To do so, we must recognize the existence of different levels of impact created by human capital in a company. Some authors (i.e., [26]), based on the model of evaluation of capacity established by Kirkpatrick [29], have added an additional level to determine the value chain for the management of human talent. This is represented in the following Figure 1.1.

Figure 1.1: Human resource management value chain. Source: Authors' elaboration.

For each of these levels, we can find different types of measurement. For levels I, II, and III, we have measures that determine the economic benefits associated with a particular program or human resource policy. When compared with the investment made in this program, the degree of contribution can be established. By means of the cost/benefit ratio or the calculation of the profitability of the program or the productivity at work, we can know their total contribution to the company. In the case of levels IV and V, in which scenario we must determine the strategic contribution of human capital, the issue becomes more complex. To this end, we are required to rely on the calculation of various indicators such as the rate of strategic alignment, the ratio of retention of key personnel, employee satisfaction, work environment, and other factors. However, an obstacle arises in that this is not a direct reflection on the effect on business value creation; what really is needed are relevant performance indicators [18].

The indicators that are closer to these functions are considered global indicators, such as the rate of economic productivity of employees or the added value of human capital (human capital added value) [26]. Such indicators are criticized since individual employees do not actually make a proportional contribution to sales revenues, which may lead one to believe that the only way to improve those indicators is through reducing staff or wages [23]. However, these indicators attribute all the added value achieved to human capital, removing from consideration the role of financial capital. In this regard, it should be made clear that the value of an asset depends on its contribution of value to the company; however, in reality, the main assets generate value only in combination with other factors. For this reason, it would be a mistake to think of the value of any intangible asset as coming from outside the context of its function in combination with all other assets [1].

To summarize the preceding information, the result of the relationship between the management of human resources on the basis of human capital will have to include not only already commonly used planning and acquisition activities and programs aimed at obtaining support as well as retaining and developing human resources [30–32], but it should also measure the impact of those behaviors on the productivity and profitability of the company [33–35]. The strategic possibilities of this resource are expected to increase under this method.

Quantifying the ROI of human capital is then produced from the identification of human resource activities as a link in the structure of investment in companies, leading the discussion toward the assumption on human resources in their participation in the production system as an intangible asset [36], instead of its typical adoption as an operating expenditure.

For this reason, among the total operating expenses, the item corresponding to staff costs must be excluded. This is supported by the theory that defends human resources as intangible assets [37]. Thus, we already know the initial component of the calculation of the $\mathrm{ROI}^{\mathrm{CH}}$. Operating costs would be obtained by subtracting staff costs from the total expenditures in the accounting system. These costs commonly include the total number of payroll items and social benefits, contributions to health and retirement plans, and costs of training, development, safety, satisfaction, confidence, and participation in employee programs [38].

Since everything relating to personnel is considered an investment, it is essential to contemplate the possibility of treating it as an intangible asset. The theory behind the concept of return on assets [39] locates the accounting base of investment in human capital in the denominator of the relationship, giving rise to the return on investment (ROI) in human capital, much as one would do with any other tangible asset:

$$\mathrm{ROI}^{\mathrm{CH}} = \frac{\text{Income} - (\text{Op. Expenses} - \text{W. Costs})}{\text{W. Costs}}$$

Equation (1.1): Return on investment in human capital (source: [40]).

Although the preceding equation arises from a theoretical analysis and evidence collected by Fintz-Enz over the last 15 years in a consulting practice, there is no way to dismiss the relevance of this model since it mandatorily responds to the requirements implicit in the organizational goal of profitability, as demonstrated in what follows.

It should be noted that, from a mathematical perspective, as an indicator of the ROI, it can only increase at the expense of a decline in the denominator, implying in this case a reduction in the implementation of the budget for investment in human resources. This may also be done by an increase in income items or a decrease in other operational costs. Nonetheless, it is necessary to consider that the investments and their corresponding returns are dependent on the behavior of the market and economic conditions, apart from human behavior.

Once the analysis that theoretically shows the interaction between investment in human capital and the factors that affect the generation of business value is complete, so that we have the tools to measure it, we put forth our first working hypothesis:

H_1: *There is a positive relationship between investment in human capital and the generation of business value.*

Keeping in mind that the activity sector will significantly impact personnel costs, and considering the above stated relationship, we may establish a priori that companies that devote more economic effort to employees will be those that will make an increased investment in human capital. Therefore, such companies are more oriented toward the development of a particular service than those with a productive focus. As a result, the amount of investment in human capital will be less compared to other investments that the company makes in technology, innovation, and other areas. For this reason, we put forth our second hypothesis as follows:

H_2: *The sector of activity to which a company belongs will influence the generation of business value.*

Finally, we propose our third hypothesis, where the sector of activity of a company will moderate the relationship between investment in human capital and the creation of value in the following way:

H_3: *The sector of activity of a company moderates the relationship between investment in human capital conducted by the company and the generation of value achieved, being greater for companies in the service sector and lower for companies in the industrial sector.*

Thus, our model is shown in Figure 1.2.

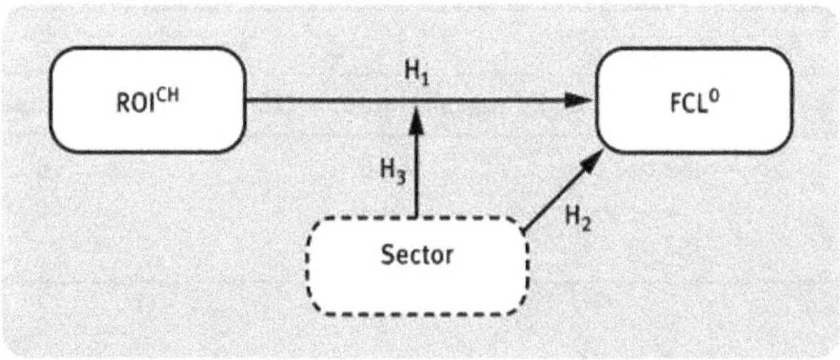

Figure 1.2: Research model. Source: Authors' elaboration.

1.3 Research design

1.3.1 Source of data, study population, and sample

To verify the hypothesis, our study population is composed of enterprises affiliated with ASCORT.

ASCORT is an association recognized in the Colombian business world as an entity that provides services in areas related to the management of human resources in the business field. Additionally, it brings together 331 important companies in the Colombian business community, including, in its database, institutions at different levels in terms of structure by size, as well as representatives of each economic sector of the country: industrial sector, financial sector, commercial sector, chemical industries sector, health services sector, public services sector, cooperatives of associated work, and official institutions. See Tables 1.1 and 1.2 and Figure 1.3 to see the composition of the sample used.

Table 1.1: Sector of activity.

Sector code					
		Frequency	**Percentage**	**Valid percentage**	**Cumulative percentage**
Valid	Education	7	30.4	30.4	30.4
	Public administration	4	17.4	17.4	47.8
	Textile manufacturing	3	13.0	13.0	60.9
	Commerce	4	17.4	17.4	78.3
	Hotels	1	4.3	4.3	82.6
	Electronic and telecom supplies	3	13.0	13.0	95.7
	Chemical manufacturing	1	4.3	4.3	100.0
	Total	23	100.0	100.0	

Table 1.2: Sectors and subsectors of activity.

Sector					
		Frequency	**Percentage**	**Valid percentage**	**Cumulative percentage**
Valid	Services	16	69.6	69.6	69.6
	Industrial	7	30.4	30.4	100.0
	Total	23	100.0	100.0	

To collect the information, we appealed directly to the enterprises, which were asked to provide information on the total value of items for 2009. The required data were obtained from balance sheets as well as the profit and loss accounts. To ensure the

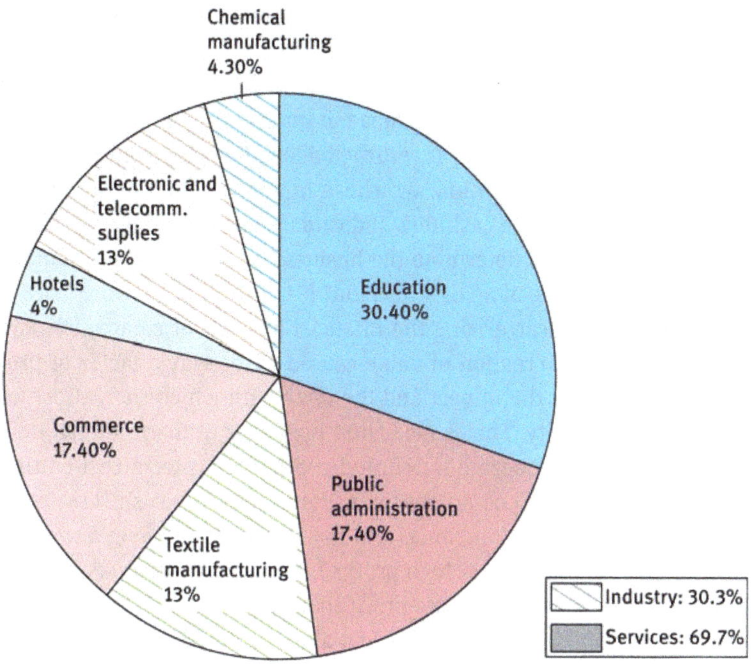

Figure 1.3: Sectors and subsectors. Source: Authors' elaboration from data of survey.

accuracy of the information, it was compared to the information contained in financial databases belonging to the chambers of commerce of Colombia.

Although the questionnaire was sent to the entire population, responses were obtained from only 23 companies, a 6.95% response rate. Though this is not a large sample of companies, the data are sufficient to achieve a minimum statistical significance through a normal distribution. In addition, this response rate is in line with that of other works using similar methodologies.

1.3.2 Study variables

To determine the profitability of the investment in human capital (ROI^{CH}) for each of the companies sampled, data on the total income for the period under study were obtained, as well as the total of operating expenses and investment in human capital, all taken from the accounting item of personnel costs.

It is worth mentioning that total personnel expenditures were complemented by the executed value of items aimed at training, welfare, and development programs, as well as the corresponding factor of social benefits. Thus, as was discussed in the theoretical analysis of the ROI, personnel costs were excluded from the total of operating costs, and they became part of the index denominator, showing that this was indeed an investment in an intangible asset [37, 41, 42].

The ROI in human capital estimated using the equation designed by Fitz-Enz [34] shows no economic amount of direct return. Instead, it indicates the number of times a unit of investment made by the company is returned depending on income.

To measure business value generation, we chose to analyze the operating free cash flow. The importance of the free cash flow indicator is based on the recognition it enjoys, making it the best tool to determine the business value or "rod of financial measure that truly shows the value of an organisation" [54, p. 2].

Given the circumstances characterizing the financial analysis of companies, any effort to show the impact on the creation of value caused of the way a factor of production is handled should justify the impact and the force with which such effort influences the generation of liquidity. This is measured by means of the free cash flow [43], given that no company can operate indefinitely with a short-term cost-cutting focus. However, if managers fail to complete their work analyzing cash flow, such that they only focus on profit, many companies will not survive in the long term [44]. We ought to consider that the presence of high profits is not necessarily related to the existence of a box to service debt or distribute profit sharing [39].

In addition, it is important to design models that allow one to shape a company's economic behavior in the ROI and future free cash flows [45]. To do so, one must take into account that the scope of corporate financial objectives may be better achieved through the use of probabilistic models, which, although at times unprecise, usually offer greater insight as compared to projects carried out blindly in an environment with so much uncertainty.

1.4 Analysis of results

As the reader can see from the results presented subsequently in Table 1.3, after contrasting the theoretical information, the minimum expected value in the ROI of human capital index equals 1, which determines the balance of the investment. In other words, if 10 euros are invested in human resources, this represents 10 euros in return for the company. A simple interpretation of this result would mean that the human resource minimally influences the generation of income, as was theoretically discussed in the section on the ROI.

To defend the participation of human capital in the generation of value in a given company, it is expected that the ROI^{CH} index will tend to be greater than 1. This indicates that the role of the human factor obviously affects revenues.

In our study, only three companies presented ROI^{CH} rates below 1, which, in the light of the theoretical framework, could be interpreted as meaning that allocating a

Table 1.3: Descriptive statistics of ROI^{CH}.

Descriptive statistics										
	N	Minimum	Maximum	Average	Standard deviation	Variance	Asymmetry	Kurtosis		
	Statistic	Statistic	Statistic	Statistic	Statistic	Statistic	Statistic	Standard error	Statistic	Standard error
ROI^{CH}	23	0.91418902	21.85218000	4.622762499	6.245244315	39.003	2.021	0.481	2.968	0.935
FCL°	23	−1599435242	1.43516E + 12	7.20600E + 10	2.98693E + 22	8.922E + 22	74.719	0.481	22.457	0.935
N valid (as per list)	23									

Source: Authors' elaboration from survey data.

budget to the human resources of these companies has a negative impact, since not even what is invested is expected to be recovered.

This is justified in accounting theory from the perspective of personnel expenditures. That is to say, obtaining a value for this index of less than 1 or even negative would always force one to contemplate the possibility that executed items, which in principle were slated for personnel projects and programs, would actually constitute an expenditure, since not even the cost of personnel is recovered.

However, it is necessary to highlight that this relates to particular cases; most of the results obtained in this study also relate to intrinsic conditions as well as to the current management implemented for each company.

If we want to know the status of a sample, Table 1.3 clearly shows that the average of this indicator for the companies analyzed exceeds four points, which leads us to assume that those companies consider their human capital to be important.

If we look at the data carefully, we see that there exists a high variability among the companies, which, at the same time, confirms the influence of each company's specific characteristics on the results.

An average of 4.60 in the results of the ROI in human resources suggests that the trend in the implementation of a human resources budget is associated with a greater proportion devoted to investment, potentially translatable into capital, in contrast to the common perspective of expenditure, as is regularly assumed in accounting systems [46].

Considering the relevance of the equation for ROI^{CH} [34], this constitutes evidence of the role played by human resources, not only through their participation as a factor of production but also in their representative role in generating business value, which ultimately converts investments into capitalizable items.

The calculation of the operating free cash flow resulted in 34.78% of the sample being negative. This does not mean, however, that the financial performance of a given company is not practical. On the one hand, it has already been established that this indicator reflects the level of liquidity of the company, such that it is highly dependent on the volume of revenue. Moreover, it is important to state that it is affected by the performance of accounts receivable and inventory management.

The preceding indicator can be used to generate for some companies, in certain cases, a margin of performance that allows them to operate under these conditions.

However, in the interpretation of these findings in aggregate form, as proposed by this study, it is possible to infer that the fact that the company has not been liquid in a given period might imply that during a given period, the financial performance of the company has failed to meet the results expected for the end of the budget year. Consequently, neither the debt service nor the distribution of profits to shareholders can be covered. In summary, these results are associated with the destruction of business value. In addition, these results may provide a guideline in the financial behavior of the company, causing biases and compromising liquidity in future periods, thus making it relevant to evaluate investment decisions.

Furthermore, one proportion of the sample study (65.22%) presents favorable results (see Figure 1.4). According to the theory of financial analysis, it would seem clear that these companies worked in a particular way with an orientation toward facilitating the creation of value for shareholders.

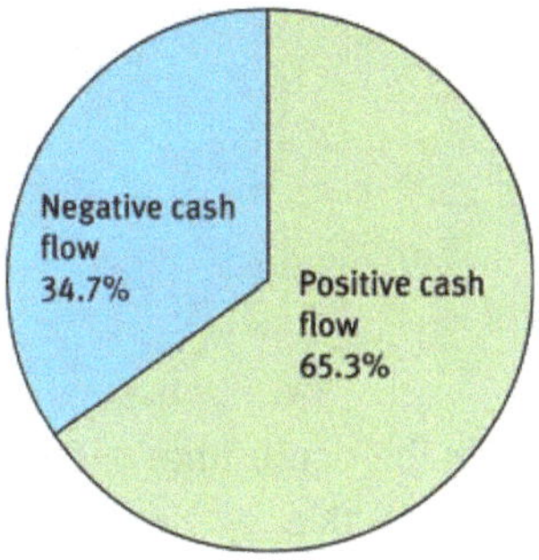

Figure 1.4: Operating cash flow: positive and negative values.
Source: Authors' elaboration from survey data.

In this analysis, it is necessary to conduct a study of correlation, which may indicate the existence of a relationship between the results of the two variables profitability of investment in human capital and business liquidity as a manifestation of the creation of value. It is also relevant to estimate the direction and strength with which they relate to each other during the verification of that relationship.

Clearly, the results must be supported by valid and reliable techniques, which is essential to estimating whether a given relationship is statistically significant.

The correlation study conducted determines the Pearson coefficient between the results of both indicators. The result indicates that the relationship between the two variables is positive, with a value of 60.0%. Such a result indicates that the relationship is direct and has a force equivalent to the percentage value (see Table 1.4). However, to achieve the objective raised by the hypothesis, this result must be interpreted in the light of its statistical significance, which, as shown in the table, must be below the *P*-value, so it should not exceed 0.05 in absolute terms. The study shows that the results are statistically significant at 99%.

From the results, it is possible to infer that there is a direct relationship between ROI in human capital, calculated using the equation of Fitz-Enz [34], and the generation of business value, deduced from the interpretation of financial behavior on the basis of operating free cash flow.

In contrast, if we consider that the sample used contains fewer than 30 data samples, it is pertinent to estimate the use of structural equations for analysis, since they present certain advantages over traditional multivariate techniques [47]. In particular, we will use the technique called PLS through the SmartPLS 3.2.3 software. This technique is suitable in the initial stages of an investigation and when few data are available, as in our case. With 5,000 subsamples obtained by using the bootstrap resampling technique, we attempt to determine the statistical significance of the coefficients in the

Table 1.4: Correlation study.

		ROI^{CH}	$FCL°$
ROI^{CH}	Pearson correlation	1	0.600*
	Sig. (bilateral)		0.002
	N	23	23
$FCL°$	Pearson correlation	0.600*	1
	Sig. (bilateral)	0.002	
	N	23	23

Note: *The correlation is significant at 0.01 (2 tails).
Source: Authors' elaboration from survey data.

structural model. This model is also examined by the value and the significance of standardized β path coefficients and the R^2 values of the dependent variable.

1.5 Results

1.5.1 Evaluation of measurement model

The individual reliability of an item need not be measured since it has a variable that has been obtained from financial statements. With respect to both the ROI^{CH} and the FCL°, the load assumes the maximum value of 1 (as shown in Table 1.5), which means it exceeds the values recommended by Carmines and Zeller [48]. Regarding the internal consistency, this is computed through composite reliability and convergent validity by applying the average variance extracted (AVE). As we predicted, these values are above those recommended, since they are 1 in both cases.

Table 1.5: Convergent reliability and validity.

FCL°	**1.000**
Moderating effect	0.963
ROI^{CH}	1.000
Sector	1.000

Finally, to measure the discriminant validity, we compare the square root of the AVE with correlations between constructs according to Fornell and Larcker's [49] recommendations. As shown in Table 1.6, the square root of the AVE is higher for all the constructs than the correlations between them, which suggest that each construct is more strongly related to its own measures than with those of others.

Table 1.6: Discriminant validity.

	FCL°	Moderating effect	ROI^{CH}	Sector
FCL°	1.000			
Moderating effect	0.662	0.981		
ROI^{CH}	0.600	0.000	1.000	
SECTOR	0.358	0.000	0.205	1.000

The first approach works well for large sample sizes; however, it is not well suited for samples of smaller scale such as ours. Therefore, to check the discriminant validity, we will analyze the HTMT criterion by intervals, which is based on a comparison of the "heterotrait–heteromethod" and the "monotrait-monomethod" correlations [50]. This will allow us to identify the discriminant validity for samples of any size. As listed in Table 1.7, none of the intervals is 1, as recommended by the existing literature [50], which again confirms the discriminant validity of the model.

Table 1.7: Discriminant validity.

	Original sample	Sample mean	Bias	5.0%	95.0%
Moderating effect	0.662	0.441	−0.221	0.000	0.506
ROI^{CH} → FCL°	0.600	0.499	−0.101	0.015	0.882
ROI^{CH} → moderating effect	0.000	0.003	0.003	0.000	0.000
SECTOR → FCL°	0.358	0.412	0.054	0.321	0.804
SECTOR → Moderating effect	0.000	0.005	0.005	0.000	0.000
SECTOR → ROI^{CH}	0.205	0.246	0.041	0.026	0.596

1.5.2 Evaluation of structural model

Once the measuring instrument has been validated, we proceed to contrast our first hypothesis, where the effect of the ROI^{CH} in the FCL° is evaluated. In Figure 1.5, we can see that the ROI^{CH} has a significant influence on FCL°, with a value of $T = 1.567$. This fact allows us to corroborate hypothesis 1 (H^1: $\beta = 0.55$, $p = 0.059$). Regardless, we cannot confirm that the sector influences in a significant manner the relationship between ROI^{CH} and FCL° because, although we find a high value for the statistic ($T = 0.966$), it does not reach the minimum value recommended to support hypothesis 2 (H^2: $\beta = 0.245$, $p = 0.167$).

To test hypothesis 3 and to measure the interaction (moderation) effect that the (industrial or service) sector establishes in the relationship between ROI^{CH} and FCL°, the effect of orthogonalization [51, 52] was examined. This method is highly recommended when few observations and few indicators are available, as in our case, and it also mitigates the potential effect of multicollinearity [52]. Although this method ac-

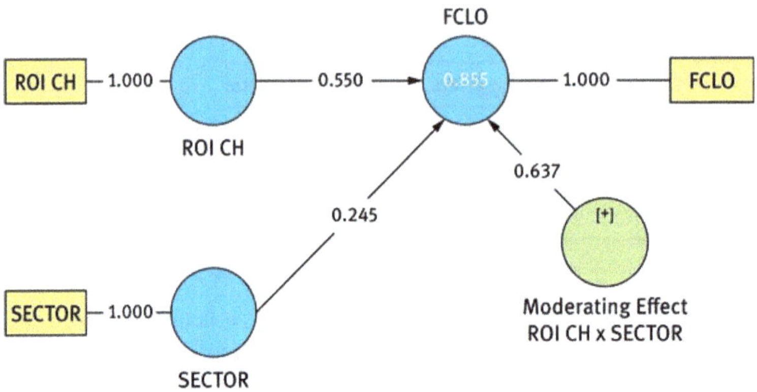

Figure 1.5: Structural model.

cepts the use of small samples, and we have $\beta = 0.637$, the significance of this effect has not yet been demonstrated since the value of the statistic T has been 0, not even achieving minimal values accepted in the literature. However, when we look at the power of the sector, we do it through the f square [53]. The value that is reached for this moderating effect is 0.498, indicating a strong moderating effect.

To better understand the strength of this interaction effect, it is represented in Figure 1.6. Even though the effect of ROI^{CH} on FCL° is extremely high in the example of the service sector, in the case of the industrial sector, this effect diminishes and even changes direction.

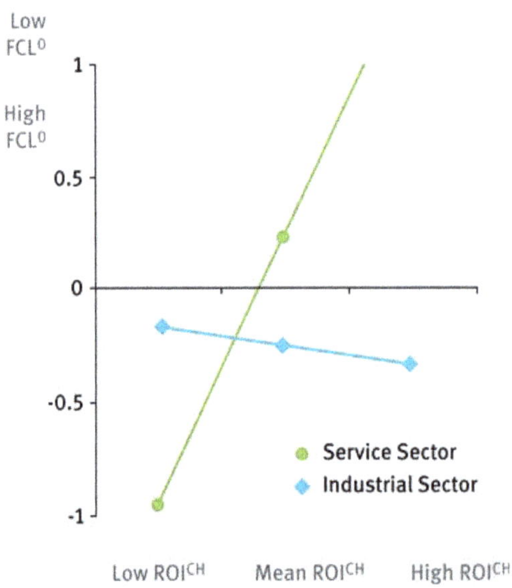

Figure 1.6: Interaction effect of FCL° and ROI^{CH} by sector.

1.6 Conclusions and discussion of results

Our work highlights the lack of economic indicators for the measurement of human capital when it comes to analyzing its contribution to the strategic development of a company. Hence the need to consider the degree of contribution of certain indicators related to investment in human capital with variables of economic and financial development of organizations is often ignored. Using the ROI^{CH} indicator, our study has led us to identify a direct and positive relationship with the generation of business value measured by operating free cash flow. Therefore, the validity of such an indicator is demonstrated when it comes to making strategic decisions since its impact on economic development is well known. Although we are aware that it is a first approximation, owing to the shortage of studies on the topic in the literature, we consider it fundamental to continue this line of work in order to establish the impact of certain global indicators of human capital. This will not only represent a step forward for the knowledge base, but it will also empower managers to use precise tools for their decision making.

We have been able to establish how ROI^{CH} can be considered a valid measurement variable to determine the contribution of human capital to the achievements of a company's strategic goals and the level of contribution of certain financial indicators such as cash flow. The sector to which the company belongs also influences this indicator.

On the other hand, and even though in the present work it has not been statistically justified, we can establish a priori that the sector of activity of a company could have a moderating effect on this relationship. Therefore, it would be necessary to deepen the study of this relationship to determine the validity of those indicators in different sectors.

Future research that may arise as a result of these findings should bring greater knowledge regarding the impact of ROI^{CH} on other economic and financial variables that companies use in their control panel, such as profitability, net profit, sales growth, market share, and so on. Getting to know the impact of a company's investment in global human capital on significant financial indicators provides a very useful management tool. Furthermore, it modifies the current view of the human factor as linked to a business cost and should be understood undeniably an asset in which the company must not only continue to investing but also expect to see profitability and business development from, in both the term and long term.

In this way, we may apply the theory of human capital to business management models, all the while remembering that the contribution made to a company by human capital occurs in combination with other resources, both technological and financial. Determining the contribution of each of the most important elements of a company to both its present and future economic development would allow for significant improvements in the senior management decision-making process, leading to a boost in competitiveness that this would bring as a result of generating more effective information systems.

References

[1] Di Francesco JM (2002). Managing human capital as a real business asset. IHIRM Journal 6:7–16.

[2] Juliano J (2007). Putting an ROI on human capital. Natural Gas & Electricity 23(9):25–28.

[3] Hudson W (1993). Intellectual Capital: How to Build It, Enhance It, Use It. New York: John Wiley.

[4] Vodak J (2010). How to measure return on investment into human capital development. Cag University Journal of Social Sciences 7(1):1–20.

[5] Colbert BA (2004). The complex resource-based view: Implications for theory and practice in strategic human resource management. Academy of Management Review 29(3):341–358.

[6] Colvin AJS, Boswell WR (2007). The problem of action and interest alignment: Beyond job requirements and incentive compensation. Human Resource Management Review 17(1):38–51.

[7] Wright PM, McMahan GC, McWilliam A (1994). Human resources and sustained competitive advantage: A resource-based perspective. International Journal of Human Resource Management 5(2):301–326.

[8] Wright PM, Dunford BB, Snell SA (2001). Human resources and the resource based view of the firm. Journal of Management 27:701–721.

[9] Arthur J (1994). Effects of human resource systems on manufacturing performance and turnover. Academy of Management Journal 37:670–687.

[10] Huselid MA (1995). The impact of human resource management practices on turnover, productivity, and corporate financial performance. Academy of Management Journal 38:635–672.

[11] Appelbaum E, Bailey T, Berg P, Kalleberg A (2000). Manufacturing Advantage: Why High-performance Work Systems Pay Off. Ithaca, NY: ILR Press.

[12] Bae J, Lawler JJ (2000). Organizational and HRM strategies in Korea: Impact on firm performance in an emerging economy. Academy of Management Journal 43:502–517.

[13] Bartel AP (2004). Human resource management and organizational performance: Evidence from retail banking. Industrial and Labour Relations Review 57:181–203.

[14] Wright PM, Gardner TM, Moynihan LM, Allen MR (2005). The relationship between HR practices and firm performance: Examining causal order. Personnel Psychology 58:409–446.

[15] Sun LY, Aryee S, Law KS (2007). High-performance human resource practices, citizenship behavior, and organizational performance: A relational perspective. Academy of Management Journal 50(3):558–577.

[16] IBM (2002). Global Human Capital Survey. Retrieved from http://www-935.ibm.com/services/us/gbs/bus/pdf/ge510-9455-global-human-capital-survey.pdf, last accessed 2016-06-26.

[17] Rigby D (2001, 13 August): Downside of downsizing. Financial Times.

[18] Pricewaterhouse Coopers (2015). 18th Annual Global CEO Survey. A new take on talent. Retrieved from http://www.pwc.com/gx/en/financial-services/publications/assets/fs-publication-a-new-take-on-talent-feb-2015.pdf, last accessed 2016-05-15.

[19] Gardner TM, Wright PM (2009). Implicit human resource management theory: A potential threat to the internal validity of human resource practice measures. The International Journal of Human Resource Management 20(1):57–74.

[20] Gerhart B (1999). Human Resource Management and Firm Performance: Measurement Issues and Their Effect on Causal and Policy Inferences. In: Research in Personnel and Human Resources

Management, Supplement 4. Wright PM, Dyer LD, Boudreau JW, Milkovich GT (eds) Greenwich, CT: JAI Press, 31–51.

[21] Rush MC, Thomas JC, Lord RG (1977). Implicit leadership theory: A potential threat to the internal validity of leader behavior questionnaires. Organizational Behavior and Human Performance 20:93–110.

[22] Wintermantel RE, Mattimore KL (1997). In the changing world of human resources: Matching measures to mission. Human Resource Management 36(3):337–342.

[23] Bukowitz WR, Williams RL, Mactas ES (2004). Human capital measurement. Research Technology Management 47:43–49.

[24] Deloitte (2015). Tendencias globales en capital humano en 2015. Liderando en el nuevo mundo. Retrieved from http://www2.deloitte.com/content/dam/Deloitte/ar/Documents/human-capital/Tendencias%20Globales%20en%20Capital%20Humano%202015.pdf, last accessed 2016-12-10.

[25] Bontis N, Fitz-Enz J (2002). Intellectual capital ROI: A causal map of human capital antecedents and consequents. Journal of Intellectual Capital 3(3):223–247.

[26] Torres JL (2005). Enfoques para la medición del impacto de la gestión del capital humano en los resultados de negocio. Pensamiento & Gestión 18:151–176.

[27] Sung T (2002). ROI of human capital. T+d 56(11):10.

[28] FASB (2006). Board Meeting Minutes 07-26-06. Retrieved from http://www.fasb.org/jsp/FASB/Document_C/DocumentPage&cid=1218220109217.

[29] Kirkpatrick D (1979). Techniques for evaluating training programs. Training and Development Journal 33(6):78–92.

[30] Fitz-Enz J (1999). ¿cómo Medir la Gestión de Recursos Humano?. Alameda Recalde: Ediciones Deusto S.A.

[31] Fitz-Enz J (2009b). How to Measure Human Capital's Impact on Processes. In: ROI of Human Capital: Measuring the Economic Value of Employee Performance. American Management Association International. New York: AMACON, 66–105.

[32] Phillips JJ, Phillips P (2009). Using ROI to demonstrate performance value in the public sector. Performance Improvement 48(4):22–28.

[33] ICON Group International, Inc. (2001). Financial Indicators: Computer Programming Services. ICON Group International, Inc.

[34] Fitz-Enz J (2003). El ROI Del Capital Humano: ¿cómo Medir El Valor Económico Del Rendimiento Del Personal?. Barcelona: Deusto.

[35] Subramony M, Krause N, Norton J, Burns GN (2008). The relationship between human resource investments and organizational performance: A firm-level examination of equilibrium theory. Journal of Applied Psychology 93(4):778–788.

[36] Ballester M, Garcia-Ayuso M, Livnat J (2003). The economic value of the r&d intangible asset. European Accounting Review 12(4):605–633.

[37] Bhattacharya M, Wright PM (2005). Managing human assets in an uncertain world: Applying real options theory to hrm. International Journal of Human Resource Management 16(6):929–948.

[38] Davoine L, Erhel C, Guergoat-Lariviere M (2008). La calidad del trabajo según los indicadores europeos y otros más. (Spanish). Revista Internacional Del Trabajo 127(2/3):183–221.

[39] García O (1999). Administración Financiera: Fundamentos Y Aplicaciones (3rd edn, vol 1). Medellín: Prensa Moderna Impresores S.A.

[40] Fitz-Enz J (2009a). How to Measure Human Resources' Value Added. In: ROI of Human Capital: Measuring the Economic Value of Employee Performance. American Management Association International. New York: AMACON, 106–143.

[41] Amit R, Schoemaker PJH (1993). Strategic assets and organizational rent. Strategic Management Journal 14(1):33–46.

[42] Bullen ML, Eyler K-A (2010). Human resource accounting and international developments: Implications for measurement of human capital. Journal of International Business & Cultural Studies 3:1–16.

[43] Huselid MA, Jackson SE, Schuler RS (1997). Technical and strategic human resources management effectiveness as determinants of firm performance. Academy of Management Journal 40(1):171–188.

[44] Alix J, Ramaekers L (1995). Cash is king. Chief Executive 104:40–44.

[45] Khosrowshahi F, Kaka AP (2007). A decision support model for construction cash flow management. Computer-Aided Civil & Infrastructure Engineering 22(7):527–539.

[46] PUC (1995). Reglamento de Normas Contables Y Plan Único de Cuentas. Santa Fe de Bogotá, D.C.: ECOE Ediciones.

[47] Haenlein M, Kaplan AM (2004). A beginner's guide to partial least squares analysis. Understanding Statistics 3(4):283–297.

[48] Carmines EG, Zeller RA (1979). Reliability and Validity Assessment. Sage University, Paper Series on Quantitative Applications in the Social Sciences. N. 7–17. Beverly Hills, CA: Sage.

[49] Fornell C, Larcker DF (1981). Evaluating structural equation models with unobservable variables and measurement error. Journal of Marketing Research 18:39–50.

[50] Henseler J, Ringle CM, Sarstedt M (2015). A new criterion for assessing discriminant validity in variance-based structurel equation modeling. Journal of the Academy of Marketing Science 43(1):115–135.

[51] Little TD, Bovaird JA, Widaman KF (2006). On the merits of orthogonalizing powered and product terms: Implications for modeling interactions among latent variables. Structural Equation Modeling 13:497–519.

[52] Henseler J, Chin WW (2010). A comparison of approaches for the analysis of interaction effects between latent variables using partial least squares path modeling. Structural Equation Modeling 17:82–109.

[53] Cohen J (1988). Statistical Power Analysis for the Behavioral Sciences (2nd edn). Hillsdale, NJ: Lawrence Erlbaum Associates, Inc.

[54] Bedoya JH (2007). 10 conceptos sobre el flujo de caja libre. Retrieved from http://www.degerencia.com/articulo/10_conceptos_sobre_el_flujo_de_caja_libre, last accessed 2016-07-21.

Ana Paula Ferreira and Marta Lopes

2 A case of certified units in a Portuguese University: interactions of ISO 9000 norms with HRM practices, employee performance, and employee satisfaction

Abstract: This study aims to assess organizational changes arising from the certification process according to the ISO 9001: 2008 norm, as well as assess the changes in human resource management (HRM) practices and in employee performance and satisfaction. From a population of employees in 3 certified service units of a Portuguese university, 248 valid questionnaires were used to support this research. Results are in line with previous research regarding general changes to the implementation and certification of ISO 9000: (1) internal restructuring – related to units' management and functioning, as well as with the employees' awareness of the quality of the services provided; (2) external restructuring – linked to the promotion of units' image/reputation and customers' satisfaction. Regarding HRM practices, some changes were identified in the planning of work, job analysis and description, and training and development practices. Some consequences in employees' work were also perceived by the respondents: data show a perception of increased individual efficiency and efficacy in employees' job performance, greater quality in orientation and control, increased levels of motivation and accountability, and increased workload. Summing up, data show changes derived from the implementation of the certification process: (1) changes in the execution and organization of work and in internal organizational procedures, (2) concerns of individuals and the organization regarding the quality of the service provided, (3) concerns with a unit's external image/reputation, and (4) increased bureaucracy. Specifically regarding employee satisfaction, this certification process did not seem to negatively affect its levels.

2.1 Introduction

Quality, certification, and its implementation through specific tools and techniques are embedded in political and economic agents' discourse, and it is also a concern of managers, employees, and consumers. It started playing an important role in companies' management after it was identified as a vital factor in policies and changes in organizations, enabling them to be more efficient and, as a result, more competitive in a global marketplace [1, 2]. Specifically, managing using a total quality management (TQM) approach seems to improve efficiency and product and service quality, reduce costs, increase customer satisfaction, and improve company performance (e.g., [2, 3]).

https://doi.org/10.1515/9783112206775-002

The ISO 9001:2008 norm, as an approach to helping organizations manage through quality principles, establishes requirements for the creation of a quality management system (QMS) that effectively addresses customers' demands through standards concerning continuous improvement, guaranteeing conformity of products or services with client demands and according to existing regulations. This management system aims to run and control an organization through quality issues. Quality, here, should be understood as the degree to which a set of inherent characteristics fulfills requirements [4]. Obtaining certification for a QMS implemented according to the ISO 9001:2008 norm means that a third party gives a written guarantee that a product, process, or service is in conformity with specified requirements [5].

Nowadays, a growing number of companies are looking to obtain ISO 9001 certification of their QMS to assure the quality of their products or services. In fact, and according to data from the 2014 *ISO Survey of Management System Standard Certifications*, 1,138,155 companies, dispersed throughout 188 countries, are ISO 9001 certified [6]. Sampaio and Saraiva [7] show that in Portugal, as in the rest of the world, QMS implementation in accordance with the ISO 9001 norm is by far the most important certification for companies. The same *ISO Survey* shows that 8006 Portuguese companies are certified according to this norm [6].

Thus, it seems that providing products or services of high quality and, as a result, being able to satisfy customer needs is a central task of all companies.

Although existing research relating QMS implementation to the role of human resource management (HRM) practices is scarce, the existing evidence seems to point to an important link between these practices (e.g., [1–3]). Also, the numbers displayed by the ISO surveys show the importance of the QMS implemented according to the ISO 9000 and 9001 norms. As a result, studies linking this particular set of quality certification processes and HRM seem to be important in helping organizations to improve the process and the expected results of its implementation.

Because those studies seem to indicate that the relationship between these dimensions has a positive effect on organizations, the present study aims to contribute to enhancing knowledge on the subject. Specifically, it aims to assess organizational changes that arose from the implementation of a quality certification process using ISO 9001, as well as the interactions with HRM practices and with employees' performance and satisfaction. The target population was the already certified service units – the academic service unit, the social service unit, and the documentation service unit – of a Portuguese university, the University of Minho (UM). The data used here came from a broader case study on the perceptions of managers and employees regarding QMS implementation [8].

This chapter starts by defining the dominant concepts at hand: quality and QMSs, the context of ISO 9000 and 9001 norms, HRM practices, and the relationship between QMS and HRM. Then the methodology used for the study is presented, along with a discussion of the results. Finally, some recommendations are offered.

2.2 Quality in the management of organizations

The common use of *quality* as a synonym for *attribute* or *characteristic* does not make its definition any easier. Even recognizing when it is lacking, quality is still subjective, complex, multifaceted, and difficult to accurately measure. Top authors define quality quite differently: Juran [9] states that quality means "fitness for use" because products and services must incorporate customers' viewpoints; Crosby [10] asserts that it is conformity to requirements; Deming [11] defines quality as anything that improves the product in customers' eyes, and Tribus [12] says it is what makes customers feel passion for a product or service.

In a management context, quality is considered a philosophy and culture that puts clients at the core of an organization's activity, conveying the message that the entire organization exists to satisfy the client. In fact, quality as a new management paradigm is present in the work of Philip B. Crosby (1926–2001), William E. Deming (1900–1993), and Joseph M. Juran (1904–2008), and they treat quality as a useful and differentiated management tool. Quality can also be seen as a process to be built around a group of norms that provide a set of principles, steps, and recommendations to help managers ensure organizations' success. This perspective evolves from quality control tools into quality management and, ultimately, TQM (e.g., [13, 14]).

The norms under ISO 9000 were compiled by the International Organization for Standardization (ISO) with the collaboration of a group of worldwide experts. These norms are at the core of the QMS implementation and certification of companies all over the world. Their main purposes are to ensure customer satisfaction and continuous improvement of organizations through quality management.

The quality concept also leads to the development of models of *excellence*. The Malcolm Baldrige[1] (USA, late 1980s) and the European model of EFQM[2] (early 1990s) were conceived by a small number of senior CEOs and focused on top management teams to achieve corporate excellence as a way to sustainable success.

---

1 The Malcolm Baldrige model of excellence is used in the USA to give a National Quality Award. This model considers seven key points used to assess the QMS: leadership, strategic planning; customer and market orientation; information measurement, analysis, and management; human resource orientation; process management; and results [15].

2 The European Foundation for Quality Management (EFQM) model considers an organization's (self-)assessment through three integrated components: (1) the eight fundamentals of excellence, which define principles for an organization to achieve sustained excellence; (2) the nine EFQM criteria that convert the fundamentals in a management tool, which are separated in five enablers (leadership; people; strategy; partnerships and resources; process, products, and services) and four outcomes (people, customers, society, and businesses); and (3) the RADAR – required **R**esults, plan a develop **Ap**proaches, **D**eploy approaches, **A**ssess and **R**efine approaches and deployment – a tool to lead to systematic improvement in all of an organization's areas [18].

The widespread application of ISO 9000 and the movement of business excellence, owing to the American and European quality awards, are some of the dominant approaches to managing through quality principles [16].

Thus, and despite the importance and recognition obtained by each of the "groups" of the presented models, the dimension of the ISO 9000 phenomenon, achieved largely through the ISO 9001 norm, gives relevance to this study. As a result, this research will adopt the definition of quality according to the ISO 9000: 2005 norm: the degree of satisfaction of requirements given by a set of intrinsic characteristics.

2.2.1 ISO 9000 norms

Founded in 1947, the International Organization for Standardization is an "independent, non-governmental international organization with a membership of 161 national standards bodies. Through its members, it brings together experts to share knowledge and develop voluntary, consensus-based, market relevant International Standards that support innovation and provide solutions to global challenges" (http://www.iso.org/iso/home/about.htm, retrieved 23 May 2016).

The ISO 9000 norms were born in 1987 as an expression of international consensus on good management practices (http://www.iso.org/iso/home/about/the_iso_story.htm#12, retrieved 22 June 2016). Since then, those norms have undergone three reviews (1994, 2000, 2008) to incorporate the theories of quality that have emerged. In this series, those that relate to QMSs are as follows:[3]

- NP EN ISO 9000: 2005 – covers the basic concepts and language.
- NP EN ISO 9001: 2008 – sets out the requirements of a QMS.
- NP EN ISO 9004: 2011 – focuses on how to make a QMS more efficient and effective.

These general norms of universal application "are undoubtedly the best known ISO publications and have been widely accepted as the basis for organizations to gain confidence of their customers and other interested parties, on their ability to understand customer requirements, legal requirements and regulations, and to systematically provide products and services that meet those requirements" [17, p. 16].

Certification is not an ISO requirement; however, ISO 9001 is designed to enable an organization to demonstrate compliance with it, using an independent third party, a certification body, with the intention of enhancing the confidence of current and future customers that it has the ability to consistently provide conforming products

3 These are the norms used in the present work. Some of those norms were reviewed in 2015.

[18]. Thus, associated with the 9000 series there is also ISO 19011: 2011, which depicts the guidelines for auditing management systems.

The ISO 9001: 2008 norm is the benchmark for the implementation of a QMS in organizations. The QMS requirements specified in this international standard complement requirements for products and can be used by internal and external parties, including certification bodies, to assess an organization's ability to meet customer demands and statutory and regulatory requirements that are applicable to the product and in line with the organization's own requirements [19].

Eight quality management principles described in ISO 9000: 2005 underlie the development of ISO 9001: 2008; they may be summarized as follows [4, 17]:

- Customer focus: Organizations must understand the needs of customers (current and future), satisfy their requirements, and seek to exceed their expectations.
- Leadership: Leaders must create and maintain an internal environment that allows people's full involvement in achieving organizational goals.
- Engaging people: As the essence of an organization, organizational members should be fully involved; this enables the use of skills for the benefit of the organization.
- Process approach: Managing activities and resources as a process allows organizations to obtain the desired results more efficiently.
- Managing with a system approach: To identify, understand, and manage interrelated processes as a system contributes to achieve organizations' goals more efficiently and effectively.
- Continuous improvement: Continual improvement of overall performance should be a permanent aim of an organization.
- Approach to decision making is evidence-based: Effective decisions are the result of data analysis and information.
- Mutually beneficial relationships with suppliers: A relationship of mutual benefit between the organization and its suppliers enhances the ability of both parties to create value.

The ISO 9001: 2008 norm also proposes a specific model of a QMS process based and centered on a Plan-Do-Check-Act (PDCA) methodology: *plan*, to establish the goals and necessary processes to deliver results in accordance with customer requirements and the organization's policies; *do*, to implement those processes; *check*, to monitor and measure processes and products, comparing them to the policies, objectives, and requirements for the product, and to report the outcomes; *act*, to continually improve the performance of processes [19].

Generically, the purpose of an audit of an implemented QMS according to ISO 9001: 2008 is to assess whether an organization has identified, and is managing, its processes using the PDCA methodology in order to obtain the desired results, which means "conforming products" [17].

2.2.2 Quality and human resource management

It is commonly accepted that human resources (HR) are one of the main assets of organizations and one of the factors that determine their progress [20]. HRM includes all activities that organizations dedicate to attract and retain employees and to ensure they have the highest possible performance to contribute to the achievement of an organizational purpose. These activities constitute the organization's HRM system, which is designed so that it is internally consistent, consistent with the other elements of the organizational architecture, and consistent with the strategy and objectives of the organization in order to improve efficiency, quality, innovation, and the response time to customer inquiries or complaints – the four basic elements of competitive advantage, according to Jones and George [21].

The importance given to the different practices of HRM, as key elements within quality management models and paradigms, has been quite remarkable since the beginning of the quality movement. Pioneering authors in this area emphasized some of these practices in their work. For example, Juran [9] postulated the importance of training, the creation of teams, and recognition in their methodology for improving quality; Crosby [10] made reference to the need for teamwork to improve quality; and Deming [22], who in his 14-point list, explicitly mentioned the implementation of job training, education and self-improvement programs, and employee involvement. All theoretical proposals in defining the concept of quality management that emerged later also included items directly related to HRM such as teamwork, education, training, and employee involvement [23].

Still, Vouzas [16] points out that the relationship between HRM and quality was widely underestimated. Only recently have quality experts, researchers, academics, and other professionals realized and clearly assumed that issues related to HR are at the heart of a quality philosophy and that their involvement and commitment are essential for the successful implementation of any initiatives, practices, and techniques aimed at improving quality. Additionally, it was also recognized that quality management has important implications for managing and organizing work and, in parallel, for HRM.

The ISO 9001: 2008 norm underlines, as mentioned in clause 6, the importance of "resource management." In subclause 6.1 it is stated that the provision of resources should be determined and provided by the organization. In turn, subclause 6.2, related specifically to HR, emphasizes both the key role of workers in the conformity of product requirements and the need for them to have the opportunity to have education, training, know-how, and appropriate experience. It is also the responsibility of organizations to ensure the competence, training, and awareness of its HR. In addition, subclause 5.5 ("responsibility, authority and communication") underlines the leading role of top management in defining and stating responsibilities, communicating results, and promoting awareness of QMS in the entire organization.

The aforementioned requirements demonstrate the relevance that ISO 9001 attaches to planning, skills, training, participation, and communication. At the same time, among the basic principles of this norm, the key role of people in an organization's performance is stressed, as is the relevance of their full involvement. For this reason, the importance of HR, according to Hassan [24], has often been highlighted in mission statements, policies, and quality strategy.

Nevertheless, several criticisms of the ISO 9000 movement have been addressed. In the first place, it has created an industry of millions, with a vast number of individuals and organizations depending on it for their subsistence: consultants, auditors, quality managers, companies, and training and certification agencies (and their workers) [25]. These criticisms also extend to the fact that it is a very expensive and generic process, with a heavy "red tape" component, a process that seems to be out of context in relation to HR, and is also heavily dependent on external auditors [25, 26], especially in the first versions of the norm.

In fact, very little empirical knowledge is available on this subject. The literature linking HR issues with quality and certification issues is still quite limited, especially given a focus on the relationship between those subjects and the impact of the implementation of ISO 9001. Also, most of these studies are descriptive in nature, seeking to contribute only a better understanding of the role of HRM in improving quality. Furthermore, the relationship between HRM and ISO 9000 certification is often reduced to HRM commitment in the design and implementation of a QMS [1, 3, 16].

However, Bayo-Moriones and colleagues [23] point to empirical studies that have already established a positive relationship between an ISO 9001 certified QMS and HR training and development (T&D). Effectively, integrating the QMS with HRM seems to be increasingly recognized as an important success factor. According to Hassan [24], experts, researchers, and quality professionals argue that people are at the center of the quality philosophy, and they argue that unless the involvement and commitment of people is achieved, the success of any quality improvement program will be compromised.

Quazi and Jacobs [26] studied the nature and extension of the impact of ISO 9001 on T&D in firms in Singapore and found that ISO 9000 certification affected training needs, models, methods, and evaluation. Improvement was reported in dimensions like HR development, hours of training, and types of training.

Hassan et al. [27] developed a study aimed at measuring employees' perceptions of HR development practices. The idea was to explore whether ISO certification led to improvements in HR development systems, examine the role of HR development practices on workers' personal development, and assess quality orientation in organizations. Participants in the study included 239 workers belonging to 8 organizations (4 of them with ISO certification), who responded to a questionnaire assessing the following systems: career, job planning, development, self-renewal, and HR development. Results showed a higher average score of ISO-certified companies compared to non-ISO-certified ones. Regarding the issues related to ISO certification and to orientation toward quality, this study concluded that companies with ISO certification had significantly higher ratings concerning career systems, which included HR planning, recruitment, performance evaluation systems, and promotions, as well as the level of job planning and quality orientation. In the remaining variables, the mean differences were not significant, although so-called ISO companies always enjoyed some advantages over the non-ISO ones.

Rodríguez-Antón and Alonso-Almeida [28] conducted a study on QMS certification and its impact on job satisfaction in services with high customer interaction. According to the authors, many organizations have implemented a QMS to increase their competitiveness, which allowed them to systematize and improve internal processes, as well as see organizational and financial returns. However, the researchers noted that there was little evidence on the impact of ISO implementation and certification on job satisfaction. Although it seems obvious, few empirical studies have argued that the more satisfied workers are, the greater their commitment to the organization, the better they serve clients, or the more their behavior influences the results derived from quality management. Furthermore, the same researchers found some studies showing that more highly skilled workers offer better customer service, which contributes to customer satisfaction, and organizations with a QMS that leads to improvements in working conditions promote greater involvement of workers, increased motivation, and a feeling of safety at work.

So in order to assess these claims, Rodríguez-Antón and Alonso-Almeida [28] developed and administered a questionnaire of managers of 943 ISO 9000-certified Spanish hotels. Data showed that this sample did not corroborate the conclusions drawn in the earlier literature, perhaps because the workers were not directly accessed.

Prates and Caraschi [29] studied the organizational impact of ISO 9001 implementation and certification in Brazilian paperboard companies. From their literature review they also concluded that the HR area is one with greater participation in this process, especially regarding worker T&D, given that the HR area is responsible for developing, implementing, and managing policies and procedures of T&D to attract and maintain employees' involvement regarding the organization's QMS. Moreover, it is essential that workers realize that job descriptions, training plans, and the evaluation of training effectiveness are directly interconnected, and therefore skill development and an appropriate task distribution facilitate the production process, reduce failures and waste, and, hence, reduce costs. While administering a questionnaire to estimate the direct impact of ISO 9001 on HR performance – by assessing dimensions like changes in work responsibility resulting from ISO 9001 certification, the role of HR in a certified QMS, the value given to the norm's principles and to customer requirements in HR T&D – these researchers discovered that ISO 9000 certification had a positive and significant impact on production, purchasing, and marketing, as well as on HR.

2.3 Certified units of Minho University

The present research was carried out on the certified service units of UM. A brief description of those units and of the university follows.

Founded in 1973, UM is a (northern) Portuguese public institute of higher education, with two main campuses, 25 km from each other. The university's teaching activities date back to 1975/1976. According to the information in the UM's activity report from 2014, UM had 18,330 students enrolled in 220 degree courses, in addition to 55 students enrolled in specialized training courses and 22 in advanced stages of scientific/postdoctoral work. Currently, the university's mission is to "produce, disseminate and apply knowledge based on freedom of thought and plural critical judgments, promoting higher education and contributing to the construction of a society based on humanistic principles – with knowledge, creativity and innovation as growth factors, sustainable development, welfare, and solidarity" [30].

UM adopts an organizational matrix model that promotes interaction among its units, with the view to carrying out projects that fulfill its mission and efficiently use its resources. The main structures of the university are its 11 teaching and research units: School of Architecture, School of Sciences, School of Health Sciences, Law School, School of Economics and Management, Engineering School, School of Psychology, Nursing School, Institute of Social Sciences, Institute of Education, and Institute of Languages and Human Sciences.

Also, UM has specific research units, cultural units, differentiated units, service units, and support offices. The last two constitute the set of infrastructures and support services for students and teachers who seek to meet the various needs of students. Among the service units are the academic services (ASU), social services (SSU), and documentation services (DSU), the only UM units that currently have certification of its QMS based on NP EN ISO 9001: 2008.

Regarding human resources, again according to the 2014 activity report [31], UM has 1,286 teachers and researchers and 602 non-teacher/researcher employees.

2.3.1 Academic services unit

According to the internal UM's regulation regarding service units [32], the ASU's purpose (mission) is the administrative management of students' educational processes, while attending to the needs of students, teachers, and the general public.

In this context, the ASU comprises divisions and sectors with different activities, including the initial training division, the graduate division, the pedagogical division, the attendance and quality division, and the secretarial service, offering their own facilities on both campuses. The ASU has 39 employees in various positions and professional careers, as well as one trainee under the UM internship program.

The QMS implemented here (according to ISO 9001: 2008) was certified in July 2012 by the Portuguese Certification Association (APCER), and this specific unit provides pedagogical and administrative support services to the teaching projects of UM.

2.3.2 Social services unit

According to the organic regulation [31], the SSU has administrative and financial autonomy and comprises five departments: Office Administrator, Administrative and Financial Department, Food Department, Cultural and Sports Department, and Department of Social Support.

The SSU's is charged with providing social support for students, including helping them with accommodations, meals, scholarships, medical and psychological support, and support for sports and cultural activities, among others. Regarding its staff, the SS employs 232 people.

In 2009 [33], this unit obtained a double certification from APCER, both for its QMS and for food safety based, respectively, on NP EN ISO 9001: 2008 and 22000: 2005. The certification of the QMS emphasizes that SSU provides services regarding meals in canteens and bars, accommodations, healthcare services, and sports and cultural activities. It is also responsible for the allocation of scholarships.

2.3.3 Documentation services unit

The DSU is, as stated in the organic regulation of the UM service units [31], an integrated system that encompasses all functional units of library and bibliographic information and all academic libraries of the university. This unit is responsible for collecting, managing, and providing, to all university sectors, scientific, technical, and cultural information necessary for accomplishing its duties and for participating in systems or networks of bibliographic, scientific, and technical information in accordance with the interests of the UM.

Relying on 39 employees to achieve its goals, the DSU is divided into the Library Division, the Information Division, and the Secretariat Section.

The DSU has implemented its QMS for the provision of documentation services, access to information, and infrastructure for research and study in Braga (the university's general library and it Congregados library) and Guimarães (Guimarães Library). APCER certification happened in July 2009.

2.4 Methods

As already mentioned, the present research aims to identify organizational changes resulting from the implementation and certification process, both in HRM and in employee performance and satisfaction.

Although there is no unanimous consensus on its validity, job satisfaction is one of the commonly used variables for understanding and assessing employee performance. Conceptually, Locke in 1976 (cited by Judge and Klinger [34]) defined job satisfaction as a positive emotional state or a state of pleasure resulting from the assessment of one's work or work experience. Thus, job satisfaction is a feeling or emotional state resulting from working conditions that will influence individuals' attitudes toward their employers. This influence can be exerted positively, benefiting organizations, or negatively, harming them.

Based on these definitions, it is not surprising that job satisfaction is constantly linked to performance, and people seem to take for granted that more satisfied workers are more productive, have better performance, and show lower turnover and absenteeism rates. However, this optimism is not always real – job satisfaction can result, for example, from a perception of easy, pleasant, or enjoyable work, combined with a generous salary; in fact, some empirical studies do not corroborate this assertion about a positive association between job satisfaction and performance [34–37].

Satisfaction at work results not only from individuals' assessments of their work but also of their lives in general. Thus, it cannot be automatically assumed that this emotional response is a *sine qua non* condition that can be used to explain performance or commitment to work, or even commitment to an organization. In fact, Cunha and colleagues [36] emphasize motivation more than satisfaction as a variable that influences behaviors. However, it cannot be ignored that some studies continue to show satisfaction as a significant dimension to explain performance because it seems to clarify why people adopt behaviors that benefit organizations such as lower voluntary absenteeism and turnover.

Advocating this line of thought, Domingues [38] states that employee satisfaction contributes to organizational success since workers' feelings influence their performance. As a result, developing and maintaining high organizational performance standards depends on employee satisfaction; hence, the importance of measuring satisfaction is increasingly recognized by quality management models.

To achieve the main purposes of this study, the following research questions were defined:
- What organizational and job changes result from a QMS implementation and certification?
- What are the interactions of QMS implementation and certification with HRM?
- What are the implications for employee satisfaction and QMS implementation and certification?

2.4.1 Procedures

A questionnaire was developed and administered to the employees of three UM service units to assess their perceptions regarding the consequences of QMS implementation and certification.[4] The questionnaire was comprised of three sections. Section I was devoted to the implementation and certification process of the QMS according to ISO 9001 and included 30 statements on which each respondent was to mark the degree of agreement on a 6-point Likert scale (with 1 denoting strongly disagree and 6 strongly agree). Section II addressed employees' satisfaction with their job and with the certification process, with 15 statements on which each respondent again indicated the degree of their satisfaction on a 6-point Likert scale (with 1 indicating very dissatisfied and 6 indicating completely satisfied). Section III was concerned with personal and professional information. Data were collected using optical recognition software, and for data analysis IBM SPSS Statistics 22.0 for Windows was used.

2.4.2 Sample

During June 2014, the questionnaires were distributed by an intermediary in each unit. Anonymity was assured, and the study's purposes were explained to participants, and the interest of the units in the study's results was also emphasized, given that the study would help them understand how to overcome a flaw in the certification process: the exclusion of workers' perceptions regarding the certification process.

The actual sample refers to the units' populations. Out of a total of 311 workers, 248 (79.7%) questionnaires were validated, where 31 (12.5%) were from the ASU, 183 (73.8%) from the SSU, and 34 (13.7%) came from the DSU.

Women made up 66.4% of the sample, and 82.1% of respondents had been working in their respective units for more than 7 years. Most of them (44.3%) had 9 years of schooling, or between 10 and 12 (32.9%), and as a result, the main professional careers were operational assistant[5] (56.9%) and technical assistant[6] (25.8%). In addition, 3.1% of study participants were managers and 11.1% were qualified staff.

The education and career dimensions were different among the three units. In the ASU, 55.2% were graduates and 58.6% were technical assistants; in the DSU, 65.6% of the participants had between 10 and 12 years of schooling, and 78.1% were technical assistants. The SSU was the unit with the lowest levels of education: 59.3% had less than 9 years of schooling, and 78% were operational assistants.

Dividing the workers from each service into those who went through the implementation and certification process and those who did not, data show that only 3.2% of ASU workers, 9.6% of SSU employees, and 6.3% of DSU workers did not experience that process.

4 The complete study [8] also involved document research and interviews with the unit service managers. For the purposes of the present work, only data regarding employees' perceptions were used.

5 Compulsory education required.

6 Required qualifications: currently, 12 years of schooling; up to 2008, 11 years of schooling.

2.5 Results

In relation to the questions about changes brought by the implementation and certification process (Section I), higher scores were related to improvements in the unit's external image, an increasing concern of the unit with customer satisfaction, increasing customer satisfaction with the unit, a more rigorous planning of work, and greater formalization of procedures. The average scores for these items were very close to 5 (agree), reflecting an actual agreement. Also, with values between 4 and 4.50, respondents showed that QMS implementation affected their own performance (e.g., "increased my orientation toward quality" or "improved my performance").

At the opposite extreme, lower average scores were obtained in connection with statements related to resistance to change, task/work flexibility and simplification, and increased productivity. Table 2.1 presents the means and standard deviations regarding the implications of QMS implementation and certification.

Table 2.2 presents the means of the five highest and lowest scored statements of each unit regarding the implications of QMS implementation and certification. Comparing the five highest scores of each unit, it turns out that the most significant items did not fully coincide. It can be seen that the units only have in common an increased concern with customer satisfaction. The ASU and DSU have three more items in common, all related to internal procedures: the amendment, clarification, and greater formalization of procedures. The SDU still has in common with the SSU the improvement of the external image of the unit. In contrast, the two items with the lowest agreement means are common to all units.

Data from Section II, regarding employee satisfaction with work and with the certification process, showed as higher scored statements the quality of service provided to customers, the unit's customer focus, the unit's QMS, working in the unit, and cooperation within the team. The lower scored items were working conditions, supervisor recognition of work performed, training provided, workload and work quality, and sharing of the unit's objectives, goals, and results (Table 2.3).

Table 2.4 presents the means of the five highest and lowest scored statements of each unit regarding employee satisfaction with work and with the certification process. Comparing, once again, the five highest scores for each service, the four most significant items are the same among the units and have the same global mean score: satisfaction with the quality of service provided to customers, focus on the customer, the QMS of the unit, and working in the unit. Also, in line with the global scores, SSU workers showed a high level of satisfaction with "cooperation with the team." ASU workers also expressed satisfaction with actions that were taken in the certification process to improve the quality of provided services, and DSU workers emphasized satisfaction with access to information needed to perform their jobs.

The two items with the lowest satisfaction scores matched among the units and the overall result. However, contrary to the global results, in the ASU and the DSU, those items are close to point 3 ("neither satisfied nor dissatisfied"); this seems to sug-

Table 2.1: Implications of QMS implementation and certification.

Item		Mean	Std. dev.
09	**Improved external image of unit**	**4.84**	**0.967**
07	**Increased unit's concern for customer satisfaction**	**4.80**	**1.035**
08	**Helped increase customer satisfaction with unit**	**4.73**	**0.953**
03	**Increased formalization of internal procedures**	**4.62**	**1.007**
05	**Work planning became more rigorous**	**4.62**	**1.097**
12	Added bureaucracy to my work	4.57	1.149
01	Changed internal procedures	4.55	1.099
02	Clarified internal procedures	4.54	1.056
04	Improved work organization	4.47	1.123
10	Improved functioning of unit	4.46	0.908
11	Increased control over my work by supervisor	4.45	1.194
25	Increased my orientation toward quality	4.45	1.238
26	Contributed to realizing the importance of my work for unit goals	4.35	1.223
16	Increased unit's concern for training issues	4.34	1.090
06	Improved communication between units	4.27	1.016
24	Increased my concern with the quality of my work	4.27	1.354
29	Increased my participation in improvement activities/tasks	4.15	1.322
22	Increased internal cooperation between units	4.14	1.170
30	The pride I feel for working in this unit increased	4.13	1.450
15	I had access to more training opportunities	4.11	1.271
19	Improved my performance	4.05	1.387
20	Increased my knowledge of the work	4.00	1.384
17	Changed some criteria regarding my performance evaluation	3.98	1.162
21	Increased my autonomy in executing work	3.94	1.314
18	Increased my commitment to work execution	3.93	1.395
14	**Increased my productivity**	**3.91**	**1.301**
13	**Simplified execution of my work**	**3.79**	**1.254**
23	**Task execution became more flexible**	**3.79**	**1.183**
27	**Gave me some concerns about changes introduced**	**3.58**	**1.319**
28	**I had some resistance to the changes introduced**	**3.18**	**1.413**

gest a certain indifference regarding "supervisor's recognition of the work performed" and "physical working conditions."

To reduce the number of variables, a principal component analysis (PCA) with Varimax rotation was performed. The dimensions found by this procedure with a minimum value of 0.60 in Cronbach's alpha (internal consistency analysis) were accepted [38].

Table 2.5 shows the PCA results regarding employee perceptions about the implications of QMS implementation and certification (Section I). Data indicate six dimensions that explain 68.6% of the total variance: **job execution and performance (C1):** reflects changes in work conditions and in the way of carrying out one's work as well as employee perceptions regarding their own performance; **employee awareness (C2):** depicts employee awareness of individual changes according to the new demands; **work procedures and organization (C3):** reflects modifications in internal

Table 2.2: Implications of QMS implementation and certification: comparison between units.

Unit	Higher	Lower
ASU	Increased the unit's concern for customer satisfaction (4.94)* Clarified internal procedures (4.90) Increased formalization of internal procedures (4.74)* Changed internal procedures (4.71) Improved work organization (4.52)	Improved my performance (3.26) Increased my commitment to work execution (3.26) Changed some criteria regarding my performance evaluation (3.23) Gave me some concerns about changes introduced (2.67)* I had some resistance to changes introduced (2.43)*
SSU	Improved external image of unit (4.92)* Helped increase customer satisfaction with unit (4.77)* Increased unit's concern for customer satisfaction (4.73)* Work planning became more rigorous (4.70)* Added bureaucracy to my work (4.66)	Increased my commitment to work execution (4.10) Task execution became more flexible (3.90)* Simplified execution of my work (3.86)* Gave me some concerns about changes to be introduced (3.81)* I had some resistance to changes introduced (3.43)*
DSU	Increased unit's concern for customer satisfaction (5.06)* Changed internal procedures (5.03) Increased formalization of internal procedures (5.00)* Clarified internal procedures (4.91) Improved external image of unit (4.85)*	Improved my performance (3.48) Increased my commitment to work execution (3.47) Increased my productivity (3.32)* Gave me some concerns about changes introduced (3.24)* I had some resistance to changes introduced (2.59)*

Table 2.3: Employee satisfaction with work and with the certification process.

	Item	Mean	Std. dev.
16	**Quality of service provided to customer**	**4.68**	**0.932**
12	**Focus on customer**	**4.49**	**0.851**
13	**QMS**	**4.46**	**0.928**
17	**Working in unit**	**4.43**	**1.134**
05	**Cooperation within team**	**4.36**	**1.153**
04	Access to information needed to perform (internal communication)	4.27	1.046
14	Changes introduced with implementation of QMS	4.27	0.991
07	Supervisor's guidance and follow-up	4.24	1.148
11	Organization and functioning of services	4.24	0.976
06	Cooperation among unit sectors	4.19	0.993
15	Actions taken to improve service quality	4.19	1.061
01	Autonomy and responsibility in developing and executing work	4.18	1.074
10	**Sharing objectives, goals, and results**	**4.10**	**1.047**
02	**Workload and quality of work**	**4.04**	**1.028**
09	**Training provided**	**3.98**	**1.046**
08	**Supervisor's recognition of work performed**	**3.84**	**1.308**
03	**Physical working conditions (e.g., ventilation, furniture, space, temperature)**	**3.57**	**1.366**

Table 2.4: Employee satisfaction with work and with the certification process: comparison between unit services.

Unit	Higher	Lower
ASU	Quality of service provided to customer (4.39) Working in unit (4.39) Focus on customer (4.26) QMS (4.23) Actions taken to improve service quality (4.10)	Cooperation among unit sectors (3.87) Access to information needed to perform (internal communication) (3.65) Training provided (3.61) Supervisor's recognition of work performed (3.39) Physical working conditions (e.g., ventilation, furniture, space, and temperature) (2.68)
SSU	Quality of service provided to customer (4.72) Focus on customer (4.53) QMS (4.52) Cooperation within team (4.42) Working in unit (4.41)	Sharing objectives, goals, and results (4.12) Training provided (4.09) Workload and quality of work (4.02) Supervisor's recognition of work performed (3.94) Physical working conditions (e.g., ventilation, furniture, space, and temperature) (3.79)
DSU	Quality of service provided to the customer (4.74) Working in unit (4.53) Focus on customer (4.50) QMS (4.35) Access to information needed to perform (internal communication) (4.32)	Actions taken to improve service quality (4.03) Cooperation among unit sectors (3.91) Training provided (3.74) Supervisor's recognition of work performed (3.67) Physical working conditions (e.g., ventilation, furniture, space, and temperature) (3.26)

procedures and in the way work is organized; **unit performance (C4):** reveals the units' concern with providing quality services; **unit's external image (C5):** reflects changes regarding the external image of the units; and **bureaucracy (C6):** single item reflecting the increase in "red tape" procedures.

Item 28 ("I had some resistance to the changes introduced") was removed from the analysis due to a loading lower than 0.5.

Another PCA was performed concerning employee satisfaction with work and with the certification process (Section II). Results are given in Table 2.6, and the two extracted factors explained 62.3% of the total variance. **Satisfaction with the units' management and functioning (S1)** translates the employees' satisfaction with the changes arising from the certification process with the unit's current organization, functioning, and management. The second dimension, **satisfaction with working conditions (S2)**, reflects employee satisfaction with the work context.

Item 3 (physical working conditions) was removed from the analysis due to a loading lower than 0.5.

To explore the interaction between the study's variables, correlations were then determined. The associations between the six dimensions regarding the implications

Table 2.5: PCA of implications of QMS implementation and certification.

Items		Loadings					
		1	2	3	4	5	6
C1	**Job execution and performance**						
19	Improved my performance	0.872					
18	Increased my commitment to work execution	0.848					
21	Increased my autonomy in executing work	0.831					
14	Increased my productivity	0.805					
23	Task execution became more flexible	0.734					
20	Increased my knowledge of work	0.723					
13	Simplified execution of my work	0.683					
22	Increased internal cooperation between units	0.658					
17	Changed some criteria regarding my performance evaluation	0.622					
15	I had access to more training opportunities	0.579					
11	Increased control over my work by supervisor	0.567					
06	Improved communication between units	0.545					
C2	**Employee awareness**						
25	Increased my orientation toward quality		0.905				
24	Increased my concern with quality of my work		0.866				
26	Contributed to realize importance of my work for unit goals		0.822				
30	The pride I feel for working in this unit increased		0.800				
29	Increased my participation in improvement activities/ tasks		0.680				
27	Gave me some concerns about changes introduced		0.527				
C3	**Work procedures and organization**						
03	Increased formalization of internal procedures			0.787			
02	Clarified internal procedures			0.764			
01	Changed the internal procedures			0.739			
04	Improved work organization			0.706			
C4	**Unit's performance**						
16	Increased unit's concern for training issues				0.743		
07	Increased unit's concern for customer satisfaction				0.517		
05	Work planning became more rigorous				0.499		
C5	**Unit's external image**						
09	Improved external image of unit					0.722	
08	Helped to increase customer satisfaction with unit					0.576	
10	Improved functioning of unit					0.506	
C6	**Bureaucracy**						
12	Added bureaucracy to my work						0.829
Eigenvalues		10.627	3.843	2.397	1.473	1.196	1.043
Total variance explained (%)		35.4	12.8	8.0	4.9	4.0	3.5
Cronbach's alpha		**0.939**	**0.876**	**0.823**	**0.723**	**0.798**	
KMO		0.888					

Table 2.6: PCA of employee satisfaction with work and with the certification process.

Items		Loadings	
		1	2
S1	**Satisfaction with unit management and functioning**		
15	Actions taken to improve service quality	0.840	
14	Changes introduced with implementation of QMS	0.815	
11	Organization and functioning of services	0.799	
10	Sharing objectives, goals, and results	0.799	
09	Training provided	0.726	
13	QMS	0.658	
08	Supervisor's recognition of work performed	0.657	
12	Focus on customer	0.630	
17	Working in unit	0.606	
01	Autonomy and responsibility in developing and executing work	0.549	
S2	**Satisfaction with working conditions**		
05	Cooperation within team		0.838
06	Cooperation among unit sectors		0.782
07	Supervisor's guidance and follow-up		0.703
04	Access to information needed to perform (internal communication)		0.666
16	Quality of service provided to customer		0.579
02	Workload and quality of work		0.564
Eigenvalues		9.406	1.187
Total variance explained (%)		55.3	7.0
Cronbach's alpha		**0.939**	**0.864**
KMO		0.941	

of QMS implementation (C1 to C6) with demographic and professional variables of education and job tenure (years in the current unit) are listed in Table 2.7.

Table 2.7: Correlations between implications of QMS and demographic and professional variables.

Demographic and professional variables	**C1 Job exec and perf**	**C2 Employee award**	**C3 Work proc and org**	**C4 Unit performance**	**C5 Unit external image**	**C6 Bureaucracy**
Education	**−0.138***	**−0.196****	**0.202****	0.013	−0.033	−0.072
Unit tenure	−0.032	−0.035	0.019	0.021	−0.035	0.059

Notes: *Correlation significant at $p < 0.05$. **Correlation significant at $p < 0.01$.

Table 2.7 shows that only education is associated with some of the implications regarding QMS implementation and certification, negatively with job execution and performance (C1) and employee awareness (C2), and positively with work procedures and organization (C3). Thus, as employees' qualifications increase (decrease), perceptions of the implications of QMS implementation regarding job execution, performance, and employee awareness decrease (increase). On the other hand, as employees' qualifications increase (decrease), perceptions regarding the consequences of QMS implementation and certification in work procedures and organization increase (decrease). The same procedure was performed regarding employee satisfaction with work and the certification process and demographic and professional variables, but no associations were found. Table 2.8 shows the correlations between implications of QMS and employee satisfaction.

Table 2.8: Correlations between implications of QMS and employee satisfaction.

Dimensions	C1 Job exec and perf	C2 Employee award	C3 Work proc and org	C4 Unit performance	C5 Unit external image	C6 Bureaucracy
S1 with unit	**0.180****	**0.584****	**0.170****	**0.177****	**0.258****	−0.034
S2 with work condition	**0.182****	**0.483****	0.069	**0.175****	**0.226****	0.018

Notes: *Correlation significant at $p < 0.05$. **Correlation significant at $p < 0.01$.

There are several positive correlations between these dimensions. In fact, satisfaction with the units' management and functioning (S1) is associated with all dimensions regarding QMS implementation, with the exception of bureaucracy. That means that employee satisfaction regarding unit management and functioning increases (decreases) when perceptions of QMS implications also increase (decrease). The same reasoning is applied to the relationship between employee satisfaction and working conditions, and all the dimensions of QMS implications, with the exception of work procedures and organization (C3) and, again, bureaucracy (C6), where no association was observed.

In an attempt to assess the possible existence of differences between each of the units regarding the dimensions of QMS implications and of employee satisfaction, it is important to compare the global average obtained by each of these dimensions to the average obtained for each unit separately.

Regarding QMS implications, Table 2.9 shows that ASU presents a lower average of agreement in three of the six dimensions, namely, C1, C2, and C6. The SS has the highest averages in five dimensions, C1, C2, C4, C5, and C6, while the DS has the C3 factor with the highest average of agreement. All in all, the degree of agreement regarding the perception of the consequences of QMS revolve around point 4 ("partially agree"), a result that reflects significant agreement with these implications.

Table 2.9: Average of QMS implications by unit.

Unit	C1 Job exec and perf	C2 Employee award	C3 Work proc and org	C4 Unit performance	C5 Unit external image	C6 Bureaucracy
ASU	3.54	3.61	4.72	4.55	4.41	4.10
SSU	4.19	4.29	4.45	4.58	4.72	4.66
DSU	3.74	3.99	4.90	4.54	4.71	4.50
Global	4.05	4.16	4.54	4.57	4.68	4.57

Regarding employee satisfaction dimensions, Table 2.10 highlights the ASU with a lower average of satisfaction in comparison to the other units. The SSU, on the other hand, has the highest average. Again, the differences do not seem significant.

Table 2.10: Average of employee satisfaction by unit.

Unit	S1 with unit	S2 with work condition
ASU	4.01	3.99
SSU	4.27	4.35
DSU	4.15	4.27
Global	4.22	4.30

Expanding these comparisons, a one-way analysis of variance (one-way ANOVA) was performed using the Bonferroni post hoc test because of the small number of comparisons to be made and the observed frequency differences. The analysis was carried out to see the effects of the variables: Unit (ASU, SSU, and DSU) and Professional Career (unit manager – UM; senior technician – ST; technical assistant – TA; operational assistant – OA; and others – O) on the score obtained by the dimensions concerning QMS implications and on those concerning employee satisfaction. Table 2.11 shows the frequencies of each independent variable.

Table 2.11: Frequencies of independent variables (units and professional career).

Independent variable	Label	n
Unit	ASU	29
	SSU	161
	DSU	31
Professional career	UM: unit manager	7
	ST: senior technician	24
	TA: technical assistant	56
	AO: operational assistant	127
	O: other	7

Table 2.12: Comparisons between QMS implications and units.

Unit		C1 Job execution and perf		C2 Employee awar		C3 Work proc and org		C4 Unit's perform		C5 Unit's ext image		C6 Bureaucracy	
(I)	(J)	Mean dif (I – J)	Sig.	Mean dif (I – J)	Sig.	Mean dif (I – J)	Sig.	Mean dif (I – J)	Sig.	Mean dif (I – J)	Sig.	Mean dif (I – J)	Sig.
ASU	SSU	**−0.6299***	0.002	**−0.7471***	0.001	0.3614	0.083	−0.0152	1.000	−0.2202	0.451	**−0.62***	0.022
	DSU	−0.2322	0.979	−0.3962	0.387	−0.2150	0.912	−0.0159	1.000	−0.3029	0.368	−0.41	0.469
SSU	ASU	**0.6299***	0.002	**0.7471***	0.001	−0.3614	0.083	0.0152	1.000	0.2202	0.451	**0.62***	0.022
	DSU	0.3977	0.083	0.3509	0.232	**−0.5764***	0.001	−0.0007	1.000	−0.0827	1.000	0.20	1.000
DSU	ASU	0.2322	0.979	0.3962	0.387	0.2150	0.912	0.0159	1.000	0.3029	0.368	0.41	0.469
	SSU	−0.3977	0.083	−0.3509	0.232	**0.5764***	0.001	0.0007	1.000	0.0827	1.000	−0.20	1.000

Note: * The mean difference is significant at 0.05.

This statistical procedure made it possible to determine, with regard to the consequences of QMS, and as translated in Table 2.12, that for C1, C2, and C6 dimensions there are significant differences between the ASU and the SSU; SSU workers show greater agreement regarding the existence of these implications. In addition, regarding the C3 dimension there are significant differences between the SSU and the DSU, where the latter shows greater levels of agreement regarding the existence of consequences on work procedures and work organization. For the C4 and C5 factors no significant differences emerged.

Results on employee satisfaction show no significant differences between S1 and S2 factors (Table 2.13). However, testing a dimension of overall satisfaction – S global (all items of employee satisfaction were included: $\alpha = 0.948$), a significant difference between the ASU and the SSU can be seen, where SSU employees reveal a greater overall satisfaction with QMS implementation and certification.

Table 2.13: Comparisons between employee satisfaction and units.

Unit		S1 with unit		S2 with work condition		S global	
(I)	(J)	Mean dif (I – J)	Sig.	Mean dif (I – J)	Sig.	Mean dif (I – J)	Sig.
ASU	SSU	−0.2925	0.202	−0.3680	0.053	**−0.3696***	0.034
	DSU	−0.1425	1.000	−0.2713	0.508	−0.2123	0.759
SSU	ASU	0.2925	0.202	0.3680	0.053	**0.3696***	0.034
	DSU	0.1500	1.000	0.0967	1.000	0.1573	0.794
DSU	ASU	0.1425	1.000	0.2713	0.508	0.2123	0.759
	SSU	−0.1500	1.000	−0.0967	1.000	−0.1573	0.794

Note: *The mean difference is significant at 0.05.

Examining what happens with professional careers (Table 2.14), we observe that there are only significant differences for the C1 and C3 dimensions. For C1 the difference is between senior technicians and operational assistants, where the latter express a higher level of agreement for QMS implications in job execution. With the C3 dimension, work procedures and organization, the difference is relevant between technical and operational assistants, where the former had a higher score on agreement.

Turning now to employee satisfaction with the QMS process, significant differences can be seen in S1 and S global for the same careers (Table 2.15). Both satisfaction with unit management, and functioning and global satisfaction are dimensions where unit managers produced significantly higher averages than the other careers (except “others”).

Table 2.14: Comparisons between QMS implications and professional careers.

Professional career		C1 Job exec and perf		C2 Employee award		C3 Work proc and org		C4 Unit performance		C5 Unit external image		C6 Bureaucracy	
(I)	(J)	Mean dif (I – J)	Sig.	Mean dif (I – J)	Sig.	Mean dif (I – J)	Sig.	Mean dif (I – J)	Sig.	Mean dif (I – J)	Sig.	Mean dif (I – J)	Sig.
UM	**ST**	0.9638	0.149	0.6548	1.00	0.4643	0.149	0.9663	0.082	0.7242	0.269	0.71	1.00
	TA	0.5589	1.000	0.5036	1.00	0.3571	1.00	0.4940	1.00	0.4881	1.00	0.54	1.00
	OA	0.1886	1.000	0.2773	1.00	0.8422	1.00	0.7358	0.255	0.5898	0.459	0.23	1.00
	O	0.8701	0.764	0.7000	1.00	0.9286	0.764	1.0476	0.210	0.7143	0.787	1.43	0.182
ST	**UM**	−0.9638	0.149	−0.6548	1.00	−0.4643	0.149	−0.9663	0.082	−0.7242	0.269	−0.71	1.00
	TA	−0.4049	0.709	−0.1512	1.00	−0.1071	0.709	−0.4722	0.226	−0.2361	1.00	−0.17	1.00
	OA	**−0.7752***	0.002	−0.3774	0.940	0.3780	0.002	−0.2304	1.00	−0.1344	1.00	−0.48	0.572
	O	−0.0937	1.000	0.0452	1.00	0.4643	1.00	0.0813	1.00	−0.0099	1.00	0.72	1.00
TA	**UM**	−0.5589	1.000	−0.5036	1.00	−0.3571	1,000	−0.4940	1.00	−0.4881	1.00	−0.54	1.00
	ST	0.4049	0.709	0.1512	1.00	0.1071	0.709	0.4722	0.226	0.2361	1.00	0.17	1.00
	OA	−0.3704	0.123	−0.2262	1.00	**0.4851***	0.123	0.2418	0.750	0.1017	1.00	−0.31	0.916
	O	0.3112	1.000	0.1964	1.00	0.5714	1.00	0.5536	1.00	0.2262	1.00	0.89	0.486
OA	**UM**	−0.1886	1.000	−0.2773	1.00	−0.8422	1.00	−0.7358	0.255	−0.5898	0.459	−0.23	1.00
	ST	**0.7752***	0.002	0.3774	0.940	−0.3780	0.002	0.2304	1.00	0.1344	1.00	0.48	0.572
	TA	0.3704	0.123	0.2262	1.00	**−0.4851***	0.123	−0.2418	0.750	−0.1017	1.00	0.31	0.916
	O	0.6816	0.561	0.4227	1.00	0.0863	0.561	0.3118	1.00	0.1245	1.00	1.20	0.065
O	**UM**	−0.8701	0.764	−0.7000	1.00	−0.9286	0.764	−1.0476	0.210	−0.7143	0.787	−1.43	0.182
	ST	0.0937	1.000	−0.0452	1.00	−0.4643	1.00	−0.0813	1.00	0.0099	1.00	−0.72	1.00
	TA	−0.3112	1.000	−0.1964	1.00	−0.5714	1.00	−0.5536	1.00	−0.2262	1.00	−0.89	0.486
	OA	−0.6816	0.561	−0.4227	1.00	−0.0863	0.561	−0.3118	1.00	−0.1245	1.00	−1.20	0.065

Note: *The mean difference is significant at 0.05.

Table 2.15: Comparisons between employee satisfaction and professional careers.

Professional career		S1 with unit		S2 with work cond		S global	
(I)	(J)	Mean dif (I – J)	Sig.	Mean dif (I – J)	Sig.	Mean dif (I – J)	Sig.
UM	ST	**1.1722***	0.007	0.7008	0.334	**0.9545***	0.022
	TA	**1.2391***	0.001	0.7821	0.112	**1.0802***	0.002
	OA	**1.3025***	0.000	0.7803	0.090	**1.1060***	0.001
	O	1.1413	0.073	0.6810	0.960	0.9608	0.130
ST	UM	**−1.1722***	0.007	−0.7008	0.334	**−0.9545***	0.022
	TA	0.0669	1.000	0.0813	1.000	0.1258	1.000
	OA	0.1303	1.000	0.0795	1.000	0.1515	1.000
	O	−0.0310	1.000	−0.0198	1.000	0.0063	1.000
TA	UM	**−1.2391***	0.001	−0.7821	0.112	**−1.0802***	0.002
	ST	−0.0669	1.000	−0.0813	1.000	−0.1258	1.000
	OA	0.0634	1.000	−0.0019	1.000	0.0257	1.000
	O	−0.0978	1.000	−0.1012	1.000	−0.1195	1.000
OA	UM	**−1.3025***	0.000	−0.7803	0.090	**−1.1060***	0.001
	ST	−0.1303	1.000	−0.0795	1.000	−0.1515	1.000
	TA	−0.0634	1.000	0.0019	1.000	−0.0257	1.000
	O	−0.1613	1.000	−0.0993	1.000	−0.1452	1.000
O	UM	−1.1413	0.073	−0.6810	0.960	−0.9608	0.130
	ST	0.0310	1.000	0.0198	1.000	−0.0063	1.000
	TA	0.0978	1.000	0.1012	1.000	0.1195	1.000
	OA	0.1613	1.000	0.0993	1.000	0.1452	1.000

Note: *The mean difference is significant at 0.05.

2.6 Discussion and conclusions

The results of the study show that modifications to units that arise from the certification process can be grouped into internal changes, which are related to the units' management and functioning and to the quality of services provided, and external changes, associated with the promotion of the units' image and concern for customer satisfaction. In fact, the main changes resulting from a QMS implementation and certification seem to be related to the organization, mainly the units, and to the job.

Regarding interactions of the QMS implementation and certification on HRM, in the certified service units the changes seem to be felt largely in connection with job planning (e.g., increased formalization of internal procedures, work planning became more rigorous). Training issues were also indicated once respondents reported the units' concern with such issues and access to more training opportunities.

Specifically regarding implications for employees' work, the results showed a perception of greater efficiency and individual effectiveness in performing the job, further guidance for quality, greater control, higher levels of motivation and accountability, but also an increased workload.

These results are somewhat in line with the existing literature regarding interactions between QMS implementation and HRM. Quazi and Jacobs [26] demonstrated ISO 9001's impact on T&D. Prates and Caraschi [29] report the impact of ISO 9001 on HR – work responsibility, value given to the norm's principles and to customer requirements, and T&D. In the present study, the units' employees reported having experienced increased access to training opportunities and more hours of training, as well as an increased concern and awareness regarding the quality of service provided. Hassan et al.'s [27] study of employee perceptions of HR development practices concluded that companies with ISO certification had significantly higher ratings regarding the career system, which included HR planning, recruitment, performance evaluation system, and promotions, as well as the level of job planning and quality orientation. The current research shows similar results regarding job planning and quality orientation.

Concerning performance, research from Prates and Caraschi [29] assessing the impact of ISO 9001 on employee performance showed that this certification has a positive and significant impact in some functional areas. The units' employees reported greater efficiency and individual effectiveness in performing their jobs, as well as further guidance for quality, greater control, and higher levels of motivation and accountability, which are indicators of performance. However, it is important to clarify here that the current performance evaluation system of these employees is strictly limited to legal constraints if they are civil servants. So they may not perceive the implications of a different performance evaluation in their professional lives (e.g., career progression, increased income). As such, the response to the items related to performance can be biased by these constraints.

Data analysis also revealed the existence of negative associations between education and the consequences felt by ASU, SSU, and DSU workers on job execution and performance, employee awareness of one's own work, and work procedures and organization. In addition, there are significant differences between the ASU and the SSU in the degree of agreement regarding implications on job execution and performance, employee awareness of one's own work, and on bureaucracy, where SSU employees reveal greater awareness of these consequences. There are also significant differences between the SSU and the DSU in relation to the consequences concerning work procedures and organization, where DSU workers reveal greater awareness of them. Although the collected data do not allow a precise justification for these differences, it is believed that they are related to the different levels of workers' education and length of service when QMS was implemented, as well with the time QMS has been in place at the unit.

In fact, the SS and DS units are the ones that have been certified for a longer period of time (both in 2009). This can explain the significant differences in perceiving some QMS implications between the ASU and SSU, where the ASU went through the

QMS process more recently (2012) than the SSU. Regarding differences between the SSU and the DSU that might be explained by levels of education, the SSU has fewer educated employees compared to the other units. Correlations (Table 2.7) show an association between three dimensions of the implications of the QMS and education: regarding the work procedures and organization (C3), the dimension where the means are higher, it can be seen that the higher the education level, the more positive the perceptions of the implications of QMS.

Regarding career, data revealed significant differences between operational assistants and senior technicians in terms of the consequences of QMS on job execution and performance, with the first ones revealing greater agreement. Also, significant differences were found between the technical and operational assistants regarding consequences for work procedures and organization, with the technical assistants revealing greater awareness of the existence of these consequences. It is suggested that the differences are due to the specific characteristics of each career. In fact, in legal terms, these careers are distinguished by different functional contents and different levels of complexity that increase from the operational assistant career, through the technical assistant, and achieving the highest level of complexity at the senior technician level.

Finally, the certification implications on the units' employee satisfaction were generally positive. Furthermore, there are significant differences in overall satisfaction between the ASU and SSU, in that SSU employees seem overall more satisfied. Significant differences were also observed in satisfaction with the unit's management and functioning and in overall satisfaction regarding career paths: the unit managers seem to be more satisfied compared to other workers in specific careers. Again, even though the data do not allow for a definitive explanation of these differences, it is thought that they could be related to the different ages and maturity levels of workers regarding QMS implementation in the units. Similarly, the variances in satisfaction found in different careers are believed to be related to the different degrees of complexity and functional content.

2.6.1 Final considerations

Few studies have addressed the relationship between quality (e.g., management and certification) and HRM, employee satisfaction, and performance. This small study may help to better understand how to analyze these issues, especially in a specific work context.

While it corroborates some of the findings of previous empirical studies on the impact of the ISO 9001 implementation and certification process, this study collected some new data that may facilitate understanding of the ISO 9001 phenomenon in organizations, mitigating difficulties in implementation and enhancing the benefits of the process. The data collected, especially with regard to employee satisfaction with

ISO 9001 certification, may facilitate understanding of HRM design and intervention to improve individual performance and employee satisfaction. The levels of education and the exposure time to a QMS certification might be variables worth exploring in future cases.

This work was based on a case study, and its results are not easily generalized. However, from a management perspective of trying to do whatever is possible with existing resources, this study can inspire the design of specific organizational interventions in this area.

References

[1] Wickramasinghe V (2012). Influence of total quality management on human resource management practices. International Journal of Quality & Reliability Management 29(8): 836–850.

[2] Jiménez-Jiménez D, Martinez-Costa M (2009). The performance effect of HRM and TQM: A study in Spanish organizations. International Journal of Operations & Production Management 29(12): 1266–1289.

[3] Yang C-C (2006). The impact of human resource management practices on the implementation of total quality management: An empirical study on high-tech firms. The TQM Magazine 18(2): 162–173.

[4] IPQ (2005). NP EN ISO 9000: 2005. Sistemas de Gestão da Qualidade. Fundamentos E Vocabulário Caparica: Instituto Português para a Qualidade.

[5] IPQ (2001). NP EN ISO 45020: 2001. Normalização E Atividades Correlacionadas: Vocabulário Geral Caparica: Instituto Português para a Qualidade.

[6] ISO (2014). The ISO Survey of Management System Standard Certifications. http://www.iso.org/iso/iso-survey, retrieved in October 14, 2015.

[7] Sampaio P, Saraiva P (2011). Barómetro da certificação. um retrato da certificação de sistemas de gestão em portugal. Qualidade , XL (1), 32–39.

[8] Lopes M (2015). Certificação ISO 9001 E Gestão de Recursos Humanos: Precedentes E Consequentes. Estudo de Caso Das Unidades de Serviços Certificadas da Universidade Do Minho, Master Thesis in HRM, University of Minho, School of Economics and Management. http://hdl.handle.net/1822/40975.

[9] Juran J (1974). Quality Control Handbook NY: McGraw-Hill.

[10] Crosby P (1979). Quality Is Free: The Art of Making Quality Certain NY: McGraw-Hill.

[11] Deming W (1994). The New Economics for Industry, Government, Education Cambridge: Massachusetts Institute of Technology.

[12] Tribus M (1993). Quality management in education. The Journal of Quality and Participation 16(1): 12.

[13] Martínez-Lorente AR, Dewhurst F, Dale BG (1998). Total quality management: origins and evolution of the term. The TQM Magazine 10(5): 378–386.

[14] Yong J, Wilkinson A (2002). The long winding road: The evolution of quality anagement. Total Quality Management 13(1): 101–121.

[15] AEC (2013). Malcolm Baldrige. http://www.aec.es/web/guest/%20centro-conocimiento/malcolm-baldrige, retrieved in November 14, 2014.

[16] Vouzas F (2007). Investigating the human resources context and content on TQM, business excellence and ISO 9001: 2000. Measuring Business Excellence 11(3): 21–29.

[17] APCER (2010). Guia Interpretativo NP EN ISO 9001: 2008 Leça da Palmeira: Associação Portuguesa de Certificação.

[18] EFQM (s/d). The EFQM Excellence Model. http://www.efqm.org/the-efqm-excellence-model, retreived in November 14, 2014.
[19] IPQ (2008). NP EN ISO 9001: 2008. Sistemas de Gestão da Qualidade – Requisitos Caparica: Instituto Português para a Qualidade.
[20] Bayo-Moriones A, Merino-díaz-de-cerio J (2002). Human resource management, strategy and operational performance in the spanish manufacturing industry. Management 5(3): 175–199.
[21] Jones G, George J (2010). Essentials of Contemporary Management Ryerson: McGraw-Hill.
[22] Deming W (1992). Out of the Crisis Cambridge: Massachusetts Institute of Technology.
[23] Bayo-Moriones A, Merino-Díaz-de-Cerio J, Escamilla-de-león S, Selvam R (2010). The impact of ISO 9000 and EFQM on the use of flexible work practices. International Journal of Production Economics 130(2011): 33–42.
[24] Hassan A (2009). ISO certification and human resource management. The Indian Journal of Labour Economics 52(2): 285–301.
[25] Douglas A, Coleman S, Oddy R (2003). The case for ISO 9000. The TQM Magazine 15(3): 316–324.
[26] Quazi H, Jacobs R (2004). Impact of ISO 9000 certification on training and development activities. International Journal of Quality & Reliability Management 21(5): 497–517.
[27] Hassan A, Hashim J, Ismail A (2006). Human resource development practices as determinant of HRD climate and quality orientation. Journal of European Industrial Training 30(1): 4–18.
[28] Rodríguez-Antón J, Alonso-Almeida M (2011). Quality certification systems and their impact on employee satisfaction in services with high levels of customer contact. Total Quality Management & Business Excellence 22(2): 145–157.
[29] Prates G, Caraschi J (2014). Organizational impacts due to ISO 9001 certified implementation on Brazilians cardboard companies. International Journal of Academic Research in Business and Social Sciences 4(5): 500–513.
[30] Universidade do Minho (s/d). Informação institucional. http://www.uminho.pt/uminho/informacao-institucional, retrieved in January 04, 2014.
[31] Universidade do Minho (2014). Relatório de Atividades e Contas Individuais 2014. http://www.uminho.pt/docs/relat%c3%b3rios-de-actividade/2015/06/11/uminho_2014_indi%20vidual_web.pdf, retrieved on October 20, 2015.
[32] Universidade do Minho (2010). Regulamento Orgânico das Unidades de Serviços da Universidade do Minho (aprovado pelo Despacho RT-49/2010, de 26 de abril). https://intranet.uminho.pt/Pages/Documents.aspx?Area=Despachos,%20Circulares%20e%20Delibera%C3%A7%C3%B5es, retrieved in January 04, 2014.
[33] Universidade do Minho (2009). Regulamento Orgânico dos Serviços de Ação Social da Universidade do Minho (aprovado pelo Despacho RT-46/2009, de 31 de julho). https://intranet.uminho.pt/Pages/Documents.aspx?Area=Despachos,%20Circulares%20e%20Delibera%C3%A7%C3%B5e, retrieved in January 04, 2014.
[34] Judge TA, Klinger R (2008). Job Satisfaction: Subjective Well-Being at Work. In: The Science of Subjective Well Being, Eid M, Larsen R (eds) NY:Guilford press, 393–413.
[35] Bilhim J (2006). Teoria Organizacional: Estruturas E Pessoas Lisboa: Instituto Superior de Ciências Sociais e Políticas.
[36] Cunha M, Rego A, Cunha R, Cabral-Cardoso C (2007). Manual de Comportamento Organizacional E Gestão Lisbon: Editora RH.
[37] Bateman T, Snell S (2012). Management NY: McGraw-Hill.
[38] Domingues I (2009). Gestão de Recursos Humanos Em Processos de Mudança: Possibilidades E Dificuldades. In: Tecnologia, Gestão da Qualidade E Dos Recursos Humanos: Análise Sociológica, Domingues I, Neves J (eds) Ermesinde: Edições Ecopy, 16–44.

Zlatko Nedelko and Vojko Potocan

3 Management tools for supporting productivity in organizations: empirical evidence from Slovenia

Abstract: The main purpose of this chapter is to empirically examine the utilization of the most frequently used management tools among employees in Slovenian production and service organizations and outline the benefits of commonly used management tools through the lenses of improving productivity and efficiency in organizations. In that framework, the contribution broadly addresses the productivity of organizations through the lenses of efficiency. This chapter examines the level of management tool utilization in production and service organizations, in different functional areas in organizations, compares this with the utilization of tools worldwide, and outlines the potential of management tool utilization for improving productivity of organizations. This chapter also addresses the commonly used management tools, for example, customer relationship management, lean production, Six Sigma, and total quality management. Empirical results about management tool utilization reveal that in production organizations, tools designed to support the optimization of organizational working, like benchmarking, outsourcing, and total quality management, are of paramount importance, while in service organizations, priorities are partly similar, but also different. In terms of management tool usage within functional areas, some tools are surprisingly commonly used, such as mission and vision statements and knowledge management. In terms of possible improvements, employees emphasize the improvement of planning activities in organizations owing to the utilization of management tools. The sample for this study includes 342 responses from employees in Slovenian organizations. A discussion section and a section on practical implications provide suggestions on how to use management tools to further improve the productivity and efficiency of organizations.

3.1 Introduction

Nowadays the business environment requires constant improvements from organizations in order to stay competitive or increase competitiveness [1–3]. Within a framework of retaining or increasing competitive advantage, an important issue is the efficiency and productivity of organizational functioning, which represents the backbone of organizational competitive advantage [4–6].

The existence and development of (most) organizations depend on the attainment of suitable outcomes of productive work and behavior [7–10]. To attain desired re-

https://doi.org/10.1515/9783112206775-003

sults, the organization must satisfy at last two preconditions at its point of departure: it must appropriately use its available (given or potential) resources for the attainment of results, and it must attain results by meeting the demands and needs of its (internal and external) social, business, and natural environments [7, 10, 11]. The attainment of such results is far from simple in an organization, which is a complex and complicated entity in objective reality. An organization faces many challenges concerning the selection and application of suitable approaches, methods, methodologies, and techniques for attaining its goals [12–14].

Efficiency has been and is an important point of departure, objective factor, and outcome of working in (diverse) organizations. Efficiency refers to the ratio of output to input of any system [15–17]. Organizations in the current environment assure their efficiency and productivity by using suitable managerial concepts, techniques, and tools [18–25].

In recent decades a plethora of management tools have been developed that can support and improve organizational efficiency [26–29]. The most frequently used tools include, for example, lean production, business process reengineering, Six Sigma, and supply chain management and processes, as well as tools that enhance innovativeness such as collaborative innovation, in addition to customer or supplier relationship management [2, 30]. Studies in the literature discuss at length management tools, whereas the focus here is on the most widely known and used tools, like outsourcing, strategic planning, total quality management, lean production, and customer segmentation [31–33]. Those studies usually consider one, two, or several tools, while studies that consider the majority of the most frequently used tools are rare [34, 35] and deal mainly with explaining those tools and their pervasiveness. Management literature in Central and Eastern Europe, which focuses on issues about improving, optimizing, and developing businesses, is abundant [36–38], but the role and importance of management tools for business, especially their meaning for supporting and improving organizational efficiency, are very seldom considered. Few studies have outlined the link between the use of a certain tool and the benefits from using it [39].

To summarize, there is a significant lack of evidence on the use of management tools in various types of organizations, like production versus service organizations, studies on the utilization of management tools in organizational departments, and the contribution of management tools to organizational productivity and efficiency. Thus, we have on the one hand a plethora of possible management tools for improving the efficiency of organizational work, but no clear link has been established between the utilization of management tools and their contribution to the efficiency of organizational work.

Based on the foregoing outlined starting points, this chapter contributes to the existing literature at least in the following ways. This chapter reveals the state of management tool utilization in production and service organizations as well as the utilization of management tools in various departments. Next the impact of management tool utilization on organizational improvements is empirically examined. Section 3.5

provides some examples of management tool utilization and discusses thoughts in the framework of the existing literature and findings. Finally, this chapter offers suggestions on how organizations can enhance their efficiency and productivity through the utilization of currently underused management tools and exploit the rich potential that different tools offer for increasing efficiency and can contribute to the overall success of organizations.

Based on the aforementioned concepts, the chapter is organized as follows. In the literature review section, we outline key starting points for understanding efficiency and productivity, followed by a discussion of the role and importance of management tools for achieving efficiency. We conclude this section with some data about management tool utilization worldwide. Next, we outline key facts about our survey and results regarding management tool utilization among employees from various viewpoints. The discussion will address the current state of management tool utilization in organizations and employee self-assessment of the benefits of using management tools. A section on practical implications outlines suggestions about how to use management tools to increase efficiency. We conclude this chapter by pointing out some limitations of this work and future research directions.

3.2 Theoretical background

Within the framework of the theoretical background, we will first outline key concepts about efficiency and productivity, followed by an introduction of management tools and the role and importance of management tools in improving efficiency. At the end of this section a few paragraphs are dedicated to the utilization of management tools in various worldwide areas.

3.2.1 Efficiency

A suitable understanding of efficiency must be holistic and, hence, result from a consideration of the interdependence among

- efficiency as a starting point (and resulting objective/goal) under consideration,
- efficiency as a factor of working (and behavior), and
- efficiency as an outcome of working activity.

In our research of organizations, we use efficiency to more holistically identify, define, and analyze organizations and to influence them. Various known conceptualizations of efficiency differ crucially in authors' understandings and according to (their) definitions of business. Thus, they define the goals of their consideration of efficiency of organizations as, for example, economic, business, organization, management, sociological, psychological, human, technological, and production, as the context of organizational working and behavior.

Therefore, the question arises as to what efficiency (from the viewpoint of organizations) is and how to define it for our research. Up to now, the term has been used in various contexts and for different purposes and has had different meanings. Today, efficiency is one of the most widespread (and frequently used) terms in the modern world. Yet its meaning is not essentially unified.

In modern literature, many diverse definitions of the term efficiency may be found. For example, *Webster's Dictionary* [17] gives 14 different definitions of (and more than 50 synonyms for) efficiency. Some of them are related to different viewpoints on the treatment of organizations. The definitions include the following:

- the ratio of the output to the input of any system [15, 17];
- skillfulness in avoiding wasted time and effort [17];
- the quality or property of being efficient [17];
- the quality or degree to which someone or something possesses adequate skill or knowledge for the performance of a duty [16];
- the ratio, expressed as a percentage, of the output to the input of power (energy or work per unit time) [17];
- cost of inputs for each unit of output produced [40];
- the power or capacity to produce a desired result [17];
- the measure of the effectiveness with which a system operates; it is stated as the ratio of a system's work output to its work input [41];
- "efficiency is doing better what is already being done" [11];
- economic efficiency is a general term for the value assigned to a situation by some measure designed to capture the amount of waste or "friction" or other undesirable economic features present [17].

For the purpose of our research on efficiency, the different definitions may be classified into a small number of basic groups:

- efficiency as a base or starting point (e.g., as purpose or goal) of working (and behavior) of an organization or efficiency as an interest supporting activity;
- efficiency as an element of operation or behavior of an organization; and
- efficiency as a result of the work or behavior of an organization.

Efficiency as a factor of work (and behavior) is what we are studying. In doing so, we use efficiency as a starting point, and we see the resulting objective as a basis and tool having an impact on the work and behavior of the organization being considered. On the other hand, efficiency as a result provides information about the actual productive work and about the necessary measures for calibration of planned and actual situations.

Therefore, efficiency may be understood (and necessarily holistically) and defined on the basis of an adequate (synergistic) understanding and use of all the aforementioned content definitions. A more detailed definition of the term efficiency depends on the selection and use of the methodology for its consideration (e.g., the selected viewpoints, methods, and methodologies).

Why? Efficiency is an elaborate (complex and complicated), dynamic, and comprehensive phenomenon that can, in our opinion, be adequately conceived and defined only in a holistic, systemic treatment. It makes sense to analyze it within this framework as a network of all selected significant viewpoints, levels, and areas of activity, and the analysis must include all their synergies.

In the case of efficiency of business organizations, we attempt to consider the activity of a (specific) group of (profit-oriented) organizations (or people) from a dialectical system of viewpoints (e.g., organizational, management, economic, and business, in synergy); this approach should enable the requisite holism of consideration according to a given business organization's purpose and the goals of its activity.

Based on the aforementioned findings of various authors and (the presented) starting points, the efficiency of business organizations can be best defined in the broadest terms as follows. Efficiency can be described in the most general terms as a necessarily holistic measure of the operation of the considered business system, which was created because humans concluded that it was possible to attain certain desired outcomes of organizational performance from the vantage point of various constituencies directly and indirectly affected by the organization. This is stated as a general (necessarily holistic) and (as a rule) synthetic criterion that is basically aimed at the evaluation of productive work (and behavior) in the transformation system of an organization. It is due to the selected viewpoint (or, hopefully, the dialectical system of viewpoints) defining the approach, field, and level of consideration and the chosen methodology (methods and techniques) of consideration.

A discussion on the definition of efficiency triggers the issue of the relationship between the concepts of efficiency and effectiveness. They are differently understood and defined in different sciences and even within the same science. Well-known solutions differ considerably and include a large variety of definitions. For example:

- In economics and business economics, as well as in most of the natural and technological sciences, they (the practitioners) use the concept of efficiency solely to deal with the suitability of the entire operation of an organization, including the internal and external parts of the business process and environments;
- In organizational theory they (the theoreticians) make a distinction between efficiency and effectiveness, and they use both terms. So, for instance, some authors define the suitability of the entire operation of an organization being reviewed with the term efficiency, while others describe this as effectiveness.
- In management science, they (the scientists) rather strictly delimit efficiency and effectiveness.

The differences and similarities of both concepts can be (partially) objectively explained by differences in authors' understanding of the topics in the given outer, i.e., objective, reality, such as that concerning organizations, and of their relations with their environments. In the earlier phases of coping with the theoretical consideration of the organization, authors understood it as a closed system with clearly defined boundaries separating it from its environment. Under such conditions, it made sense to separately and independently consider:

- efficiency as a goal, factor, and measure of internal processes (e.g., doing things right) and
- effectiveness as a goal, factor, and measure of external processes (e.g., doing the right things).

Efficiency in the broader sense has been covered by more recent authors who view the organization as an open system and hence examine the suitability of its working activity and behavior inside and outside the organization and discern their suitability in an appropriate way, for example, by comparing the attained and planned outcomes or the expected and real ones.

On the other hand we still use the term effectiveness, seeing it as equal to efficiency in the broader sense in order to pay tribute to the scientific correctness related to findings and references of various sciences. This applies especially to sciences of organization and management, in which many authors pay much attention to the efficiency concept.

The presented definition is formulated so as to allow us to consider the efficiency of the working and behavior of the transformation process in organizations and individuals as so-called business organizations emphasizing so-called business viewpoints rather than natural or technical/technological, social, psychological, or other viewpoints in the consideration of features, events, and processes of real life.

Based on the preceding concepts, it is seen that productivity is "hidden" in different definitions of efficiency, emphasizing the relationship between physical input and output. We use both terms – efficiency and productivity – as referring to improvements in organizational operation.

3.2.2 Management tools

To use single management ideas, researchers have also developed and presented the following useful definitions and descriptions: (1) concept – as a rather comprehensive, developed, and defined basis for the consideration of an idea; (2) methodology – as an entity or closely related collection of methods, rules, and disciplinary postulates; (3) methods – as goal- and problem-ordered types of procedures; these are especially regular and systemic ways of setting and realizing goals; (4) techniques – as the manner in which technical details are treated; and (5) necessary tools (in the management literature some authors use the term instruments instead) [18–25].

A holistic consideration of management tools – as a possible (and selected) level of contemplation of management concepts – is presented in Figure 3.1.

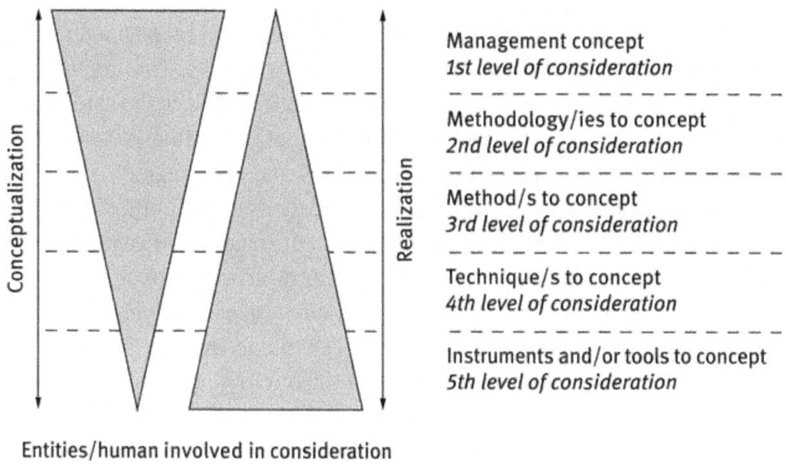

Figure 3.1: Breadth of the holistic consideration of management concepts.

A holistic comparison of several different management tools (e.g., their characteristics, preferences, weaknesses, usage, and possible results) is very complex and difficult [42–44]. In the literature, authors mention some dilemmas associated with making comparisons [20–22, 24, 45], and these authors give the following as examples of dilemmas: different comprehension and contextual understanding of tools, simultaneous usage of several tools, and various sizes and use of tools on different levels or in different fields of organizations. Not all management tools can be compared; tools that were developed for different purposes, have different aims, or are mutually exclusive depending on the context are impossible to compare.

Management tools involve a set of concepts, processes, exercises, and analytical frameworks. This definition was coined by Rigby [34] and is based on his survey of management tools. Authors from the general management literature [12, 28, 44] and other management researchers use the term management tool [46, 47], but they do not clearly define it. Building on the aforementioned findings, a management tool may be defined as an instrument that can be used to support the implementation of concepts and ideas at all levels of conceptualization and realization of concepts, with the ultimate aim of supporting organizational processes.

Rigby and his peers proposed a typology of management tools, defining four groups of management tools. A typology is framed in a matrix, having two dimensions: (1) use of management tools and (2) percentage of satisfied users. This typology classifies management tools according to the mean values of use and satisfaction obtained in research [27, 34, 35]. Owing to the lack of comprehensive studies on management tools in the literature, there is no generally valid or content-related classification of management tools.

In searching for a content-related classification, known management tools can be organized into two groups based on a historical development of concepts, current use, and their potential for use. The first group encompasses traditional management tools that were developed in earlier management development phases. The majority of such tools are currently the most well-known and most widely used tools, like benchmarking, strategic planning, mission and vision statements, customer relationship management, outsourcing, the balanced scorecard, and customer segmentation [27, 34, 35]. The literature reveals that, globally, the most widely used tools aim at supporting customer satisfaction (e.g., customer relationship management and customer segmentation) [48, 49], followed by those supporting the long-term and clear future development of an organization (e.g., strategic planning, mission, and vision statements) [47, 50, 51], supporting competitive comparison (e.g., benchmarking) [52, 53], and supporting optimization processes in an organization (outsourcing, core competencies, and business process reengineering) [32, 54, 55]. There is also evidence showing a decreasing use of some traditional tools. For example, the popularity of business process reengineering has fallen in the last two decades owing to numerous unsuccessful reorganizations of business processes [35, 47, 56]. Also, total quality management, first treated as a potential source of sustainable competitive advantage [55], became less widely used when its use did not result in sustainable profitability [35, 47].

In the second group are contemporary tools. A brand-new management concept or tool is rare since the majority of tools have their roots in earlier phases of management development [25]. This group encompasses tools whose evolution is largely based on either information technology development or supporting an existing management concept with information technology. This group also encompasses tools developed in later phases of management development. Furthermore, supporting existing management concepts with information technology results in a higher potential for this concept [25, 57]. Typical representatives of this group are, for example, shared service centers – aimed at a set of activities (e.g., human resource management activities) under one roof for selected organizational parts [58]. One of the latest tools is the corporate blog, whose serious use is a recent phenomenon [59]. Organizations use blogs for direct communication with their target population, information dissemination, or brand loyalty development, for example [59]. Further, radio frequency identification has become a widely used tool (especially in supply chains) since it enables acquiring data of any entity that can be psychically tagged and wirelessly scanned [60]. Also, relatively new are loyalty management and consumer ethnography [61]. For example, knowledge management takes on a whole new dimension when electronic databases are used to store knowledge and interfaces enable easier capture and dissemination of knowledge [57].

A detailed description of individual management tools is beyond the scope of our contribution. We will briefly describe some of the tools that will be outlined in the discussion and so will be briefly described in that section of the chapter.

3.2.3 Utilization of management tools in organizations

An overview of the utilization of commonly used tools in organizations in selected worldwide areas as well as from selected Central and East European countries, namely, Slovenia and Croatia, is given in Table 3.1 (based on [25–27, 30, 35]). Table 3.1 presents rankings of management tools in terms of utilization; the rankings are calculated based on the average utilization among employees in organizations.

Table 3.1: Management tool utilization.

Management tool[a]	Use of management tools – rank							
	Global 2006	Global 2008	North America	EU 15	Asia-Pacific	Latin America	Croatia	Slovenia
Strategic planning	1	2	1	1	2	1	8	8
Customer relationship management	2	4	3	4	1	9	4	6 (t)
Customer segmentation	3	7	6	2 (t)	3	3 (t)	5	11
Benchmarking	4	1	2	2 (t)	9 (t)	2	2	2
Mission and vision statements	5	3	4	7	5 (t)	5	1	6 (t)
Core competencies	6	9	5	5 (t)	4	10	3	3
Outsourcing	7	5	8	5 (t)	7 (t)	3 (t)	6	1
Business process reengineering	8	8	10 (t)	10 (t)	5 (t)	14 (t)	10 (t)	10
Scenario and contingency planning	9	13	9	8	14	7	10 (t)	12
Knowledge management	10	14	12	10 (t)	7 (t)	14 (t)	7	4
Strategic alliances	11	11	7	9	13	13	20	17
Balanced Scorecard	12	6	13 (t)	13	12	11 (t)	13	9
Supply chain management	13	12	13 (t)	14 (t)	11	11 (t)	12	15
Growth strategy tools	14	16	10 (t)	10 (t)	15 (t)	6	19	18
Total quality management	15	17	18 (t)	14 (t)	9 (t)	8	9	5

Table 3.1 (continued)

Management tool[a]	Use of management tools – rank							
	Global 2006	**Global 2008**	**North America**	**EU 15**	**Asia-Pacific**	**Latin America**	**Croatia**	**Slovenia**
Shared service centers	16	15	18 (t)	19	15 (t)	16	14	23 (t)
Lean operations	17	–	15	17 (t)	18 (t)	19	25	23 (t)
Collaborative innovation	18	22	16	20	18 (t)	18	17	19
Loyalty management	19	24	21	17 (t)	17	20 (t)	15	13
Mergers and acquisitions	20	10	17	14 (t)	22	17	22 (t)	14
Six Sigma	21	–	22	23	20	20 (t)	22 (t)	20
Offshoring	22	–	18 (t)	21	24	24	24	25
Consumer ethnography	23	–	23 (t)	22	21	22	18	21
Corporate blog	24	–	23 (t)	24	23	25	16	16
Radio frequency identification	25	–	25	25	25	23	21	22

[a]**Note:** Data for global average (GL) 2006 and 2008, North America (NA) 2006, European Union 15 (EU) 2006, Asia Pacific (AP) 2006, and Latin America (LA) 2006 are calculated upon results from management tools utilization research [27, 35]. Data for Slovenia and Croatia are adopted from research of management tools in Central and East Europe economies [25, 30]. Abbreviation t stands for tight result; (–) stands for not applicable.

Typical management tools aimed at supporting production organizations, like total quality management, Six Sigma, and lean production [62], are not in the forefront of use. Furthermore, the link between the utilization of different management tools and the results of their usage is somewhat blurry, even though some studies deal with the impact of the utilization of several management tools on processes or organizational outcomes.

Based on the concepts just outlined, and in line with aforementioned aim of this contribution, we will try to answer the following research questions:

- Does the level of utilization of management tools differ between production and service organizations in Slovenia?
- What is the level of management tool utilization from the perspective of departments?
- What are the key benefits of management tool utilization for increasing the productivity and efficiency of organizations?

3.3 Research design and methodology

3.3.1 Data and sample

Data for this contribution were obtained from a survey among employees in Slovenian organizations in 2014 and 2015. Based on random sampling, we sent a link to an online questionnaire to 2,000 email addresses of employees, which were obtained from company websites. A maximum five emails per organization were sent. The target population was employees at supervisory and nonsupervisory positions. We received 357 responses, resulting in a 17.85% response rate. The analysis included 342 answers since we eliminated those with more than 5% of missing data and possible indices of pattern answering.

Sample characteristics reveal that respondents are, on average, 40.31 years old, have on average 16.80 years of work experience, and have been working for their current organization for an average of 9.31 years. The sample includes 48.2% males and 51.8% females. In terms of education, 7.0% finished secondary school, 59.5% have a high school or university degree, while 33.5 have a master's degree. In terms of the current position of respondents in their organizations, 58.5% are supervisory staff (of which 10.5% is lower, 27.2% middle, and 18.1% top managers) and 41.5% are nonsupervisory staff. In terms of departments, 10.7% of the respondents work in the research and development (R&D) department, 25.0% in fundamental processes, 9.2% in accounting, 10.1% in marketing, 33.6 in a supervisory position in different departments, and 11.6 in other departments (e.g., human resources, law, IT, and logistics). Regarding size, 10.0% were organizations with less than 10 employees, 12.9% have between 10 and 49 employees, 32.9% have between 50 and 249 employees, and 44.1% have more than 250 employees. In terms of economic sector, 1.5% of the organizations operate in industries in a primary sector, 22.8% in industrial organizations in a secondary sector, 58.6% in a tertiary sector, i.e., in services, and 17.2% in a quaternary sector (e.g., nongovernmental, charity organizations, and nonprofit organizations).

3.3.2 Instrument

The instrument for surveying various aspects of management tool utilization consists of three parts. In the first part are questions related to average use of tools, in the second are questions related to implementation, benefits of utilization, and projected future use of tools, and in the third part are demographic data about respondents and organizations. The list of tools is based on the list of tools included in Bain's worldwide survey [27, 34, 35].

For measuring the utilization of a single management tool, respondents rated each tool using an interval scale ranging from "I know and use the tool" (1) to "I don't know and don't use the tool" (3). Among organizational factors, for organizational size, respondents chose one of the options from “micro” to “large” organization. Regarding their position, employees could choose from “non-supervisor staff” to “top management.” Similarly, for industry, department, and possible benefits of management tool usage, respondents chose one answer from the several quoted. Personal factors included gender, age, and years in labor force. For education, respondents could choose an option in a range from “primary school” to “Ph.D.”

3.3.3 Research design

For this contribution, we used data about management tool utilization from the first part of the research and data about benefits of management tool usage from part two. First, we outline the utilization of management tools in Slovenian organizations, where we sketch the level of management tool utilization for production and service organizations. We group organizations based on NACE classification into production (e.g., Nace from A to C) and service organizations (other remaining, excluding public services). In line with the aims of this chapter, we next emphasized the utilization of management tools in various departments in organizations, also detailed for each department. We conclude our presentation of the results by emphasizing the benefits of management tool utilization in production and service organizations, also detailed for each department.

3.4 Results

3.4.1 Utilization of management tools in production and service organizations

In Table 3.2, we present the mean values for management tool utilization, for the entire sample, and separately for production and service organizations. We also provide the respective rankings.

The results on management tool utilization indicate that Slovenian organizations commonly use benchmarking, outsourcing, mission and vision statements, core competencies, and strategic planning. A comparison with service organizations reveals that organizations involved in production use management tools to support their fundamental processes more than organizations involved in various services. The most commonly used tool among employees in production organizations is benchmarking (mean value – 1.44), while the most commonly used tool among employees in service organizations is a mission and vision statement (mean value – 1.74).

The preceding table makes it clear that some differences exist in management tool utilization between production and service organizations. For instance, benchmarking, knowledge management, core competencies, total quality management, bal-

Table 3.2: Management tool utilization in Slovenian organizations.

Management tool	All[a]	Rank[b]	Production[c]	Rank[d]	Services[e]	Rank[f]
Benchmarking	1.69	1	1.44	1	1.76	3
Outsourcing	1.71	2	1.55	2	1.75	2
Knowledge management	1.90	7	1.55	3	2.01	8
Core competencies	1.78	4	1.58	4	1.84	5
Total quality management	1.93	8	1.61	5	2.02	9
Strategic planning	1.84	5	1.62	6	1.91	6
Mission and vision statements	1.71	3	1.62	7	1.74	1
Balanced Scorecard	2.19	11	1.86	8	2.29	11
Customer relationship management	1.84	6	1.88	9	1.84	4
Supply chain management	2.24	12	1.92	10	2.33	13
Customer segmentation	1.96	9	1.94	11	1.97	7
Business process reengineering	2.17	10	1.94	12	2.25	10
Scenario and contingency planning	2.30	15	1.95	13	2.41	15
Mergers and acquisitions	2.24	13	1.95	14	2.32	12
corporate blogs	2.38	16	2.05	15	2.49	17
Strategic alliances	2.39	17	2.09	16	2.47	16
Loyalty management	2.29	14	2.12	17	2.35	14
Six Sigma	2.55	22	2.18	18	2.66	23
Shared service centers	2.46	18	2.22	19	2.53	18
Growth strategy tools	2.47	19	2.23	20	2.54	19
Collaborative innovation	2.50	20	2.29	21	2.55	20
Lean operations	2.65	24	2.40	22	2.73	24
Offshoring	2.58	23	2.44	23	2.63	22
Radio frequency identification	2.67	25	2.45	24	2.73	25
Consumer ethnography	2.55	21	2.52	25	2.56	21

Notes:

a. Mean values for utilization of management tools for all 342 cases in the analysis.
b. Ranks for management tool utilization regarding mean values of all 342 cases in analysis.
c. Mean values for management tool utilization for employees based in organizations operating in production (e.g., secondary sector).
d. Ranks for management tool utilization based on mean values for organizations operating in production.
e. Mean values for management tool utilization for employees based in organizations operating in various services (e.g., tertiary and quaternary sectors).
f. Ranks of management tool utilization based on mean values for organizations operating in various services.

anced scorecard, and supply chain management are more important in production organizations, while mission and vision statements, customer relationship management, and customer segmentation are more important in-service organizations.

3.4.2 Utilization of management tools in organizational departments

In this section, we first present frequencies of management tool utilization in organizational departments of production and service organizations. The results are summarized in Table 3.3.

Table 3.3: Utilization of management tools in organizational departments.

Department	Production	Services	Frequency
R&D	41	83	124
Accounting	42	106	148
Board of directors	44	100	144
Fundamental process	48	101	149
Marketing	51	118	169
Other	1	17	18
Total	227	525	752

Participants in the survey were asked to identify which departments in their organization use management tools. In this multiple-response question, participants could select several departments, based on their knowledge of where tools were used. Results reveal that in production organizations, management tools are most frequently used in the marketing department, followed by fundamental processes, the board of directors, accounting, and R&D. In service organizations, management tools are most frequently used in marketing, followed by accounting, the board of directors, fundamental processes, and R&D. We can see that a total of 169 respondents (51 in production organizations and 118 in service organization) named the marketing department as the department where management tools are most frequently used. Similarly, a total of 149 respondents named organizational fundamental processes as the area where management tools are most frequently used. The remaining results could be interpreted in the same manner. We can conclude that in both production and service organizations, management tools are most frequently used by the marketing department and least frequently in R&D. A word of caution is needed here since we can compare only within production or service organizations and the percentage of production organizations is significantly lower than that of service organizations.

Table 3.4: Utilization of management tools in organizational departments.

Management tool	R&D			Fundamental process			Accounting			Marketing			Board of directors			Other		
	Together	Production	Service	Together	Production	Service	Together	Production	Service	Together	Production	Service	Together	Production	Service	Together	Production	Service
Strategic planning	1.94	1.38	2.11	1.92	1.71	1.96	1.81	1.63	1.87	1.67	1.71	1.65	1.86	1.59	1.96	1.74	1.71	1.74
Customer relationship management	1.75	1.38	1.86	1.92	2.07	1.89	1.90	1.75	1.96	2.00	2.14	1.96	1.70	1.82	1.65	2.05	2.29	2.00
Customer segmentation	1.86	1.63	1.93	2.00	2.21	1.96	2.10	1.75	2.22	1.67	2.00	1.58	1.93	1.79	1.99	2.22	2.57	2.13
Benchmarking	1.72	1.25	1.86	1.80	1.79	1.80	1.81	1.38	1.96	1.56	1.57	1.56	1.62	1.33	1.74	1.66	1.43	1.71
Mission and vision statements	1.56	1.50	1.57	1.71	1.64	1.73	1.61	1.63	1.61	1.71	1.71	1.70	1.76	1.61	1.83	1.82	1.71	1.84
Core competencies	1.67	1.63	1.68	1.83	1.71	1.86	1.65	1.13	1.83	1.91	1.86	1.93	1.86	1.61	1.96	1.55	1.43	1.58
Outsourcing	1.86	1.50	1.96	1.85	1.71	1.87	1.58	1.38	1.65	1.50	1.43	1.52	1.65	1.55	1.70	1.74	1.57	1.77
Business process reengineering	2.14	1.25	2.39	2.25	2.50	2.20	1.97	1.75	2.04	2.18	2.29	2.15	2.20	1.91	2.33	2.16	1.57	2.29

(continued)

Table 3.4 (continued)

Management tool	R&D			Fundamental process			Accounting			Marketing			Board of directors			Other		
Scenario and contingency planning	2.36	1.38	2.64	2.35	2.00	2.42	2.45	2.25	2.52	2.24	1.86	2.35	2.27	2.03	2.37	2.17	1.83	2.23
Knowledge management	2.09	1.50	2.26	1.90	1.46	1.99	1.74	1.38	1.87	2.00	1.86	2.04	1.88	1.58	2.01	1.79	1.57	1.84
Strategic alliances	2.56	1.75	2.79	2.40	2.14	2.46	2.52	2.25	2.61	2.18	2.00	2.22	2.37	2.09	2.48	2.37	2.29	2.39
Balanced Scorecard	2.14	2.13	2.14	2.47	2.29	2.51	1.97	1.38	2.17	2.18	1.71	2.30	2.06	1.79	2.18	2.21	1.71	2.32
Supply chain management	2.33	1.88	2.46	2.27	1.93	2.34	2.26	1.75	2.43	2.03	1.86	2.07	2.19	1.88	2.32	2.37	2.43	2.35
Growth strategy tools	2.69	2.00	2.89	2.51	2.43	2.53	2.32	2.00	2.43	2.50	2.57	2.48	2.38	2.09	2.50	2.53	2.71	2.48
Total quality management	1.86	1.38	2.00	2.01	1.93	2.03	1.77	1.25	1.96	2.12	2.00	2.15	1.87	1.52	2.01	1.97	1.71	2.03
Shared service centers	2.44	1.88	2.61	2.44	2.36	2.46	2.65	2.50	2.70	2.65	2.57	2.67	2.39	2.09	2.51	2.43	2.29	2.47
Lean operations	2.75	2.25	2.89	2.70	2.43	2.76	2.61	2.50	2.65	2.59	2.29	2.67	2.64	2.39	2.75	2.63	2.57	2.65

Collaborative innovation	2.50	2.13	2.61	2.54	2.43	2.56	2.39	2.25	2.43	2.44	2.57	2.41	2.46	2.18	2.58	2.63	2.43	2.68
Loyalty management	2.39	2.25	2.43	2.36	2.43	2.34	2.32	2.00	2.43	2.15	2.14	2.15	2.20	1.97	2.29	2.50	2.14	2.58
Mergers and acquisitions	2.33	1.38	2.61	2.32	2.07	2.37	2.10	2.13	2.09	2.18	2.00	2.22	2.21	2.00	2.29	2.24	1.86	2.32
Six Sigma	2.53	1.75	2.75	2.60	2.43	2.64	2.65	2.38	2.74	2.50	2.29	2.56	2.47	2.15	2.60	2.68	2.00	2.84
Offshoring	2.67	2.25	2.79	2.61	2.57	2.62	2.52	2.38	2.57	2.56	2.71	2.52	2.53	2.30	2.62	2.66	2.86	2.61
Consumer ethnography	2.69	2.50	2.75	2.52	2.71	2.49	2.58	2.50	2.61	2.35	2.57	2.30	2.53	2.39	2.58	2.71	2.71	2.71
Corporate blogs	2.47	2.50	2.46	2.39	2.36	2.39	2.45	1.75	2.70	2.38	2.43	2.37	2.31	1.85	2.51	2.53	1.86	2.68
Radio frequency identification	2.58	2.75	2.54	2.81	2.86	2.80	2.58	2.13	2.74	2.76	2.86	2.74	2.55	2.21	2.70	2.76	2.43	2.84

Next, we outline the mean values for single management tool utilization in various organizational departments, for the entire sample, as well as for production and service organizations. The results are summarized in Table 3.4.

From the perspective of organizational departments, regardless of whether it is a production or service organization, the most commonly used tools in the various departments are as follows: R&D – mission and vision statements, fundamental processes – mission and vision statements, accounting – outsourcing, marketing – outsourcing, and board of directors – benchmarking. In production organizations, the most commonly used tools in the various departments are as follows: R&D – benchmarking and business process reengineering, fundamental processes – knowledge management, accounting – core competences, marketing – outsourcing, and board of directors – benchmarking. In service organizations, the most commonly used tools in the various departments are as follows: R&D – mission and vision statements, fundamental processes – mission and vision statements, accounting – mission and vision statements, marketing – outsourcing, and board of directors – customer relationship management.

3.4.3 Usage of management tools and organizational improvements

Finally, we present the results on the impact of management tools on organizational improvements. The results are summarized in Table 3.5.

In this multiple-response set respondents named areas that had been improved as a consequence of using management tools. Participants in the survey could select up to three areas of improvement. The results reveal that usage of management tools in organizations most frequently lead to improvements in planning, followed by an influence on decision-making, controlling, and leadership processes. With respect to various departments, management tools most frequently contribute to improvements in production organizations in the following areas: R&D – planning; fundamental processes – planning; accounting – planning and controlling; marketing – planning, decision-making, and controlling; board of directors – planning. In service organizations management tools most frequently contribute to the following improvements: R&D – controlling; fundamental processes – planning; accounting – planning, decision-making, and operations; marketing – leading; board of directors – planning.

3.5 Discussion

The main purpose of this chapter is to empirically examine the utilization of the most widely used management tools among employees in Slovenian production and service

Table 3.5: The impact of management tools' utilization on organizational improvements.

	R&D		Fundamental process		Accounting		Marketing		Board of directors		Other		
Field of improvement	**Prod.**	**Ser.**	**Prod.**	**Ser.**	**Prod.**	**Ser.**	**Prod.**	**Ser.**	**Prod.**	**Ser.**	**Prod.**	**Ser.**	**Total**
Planning	8	16	8	38	7	9	3	12	27	49	6	11	194
Controlling	7	18	7	36	7	7	1	13	23	40	5	8	172
Decision-making	6	16	4	34	6	9	3	14	20	42	5	14	173
Leading	3	12	4	33	6	6	1	15	23	35	5	11	154
Operations	3	11	6	28	7	9	3	13	19	36	4	10	149
Informing	3	9	2	28	5	4	1	6	18	22	5	9	112
Organizing	3	17	5	30	6	4	1	10	16	41	4	7	144
Behavior	2	7	2	23	6	4	0	7	13	25	2	7	98
Other	2	0	0	4	1	0	0	0	0	3	1	4	15
Total	37	106	38	254	51	52	13	90	159	293	37	81	

Note: *Abbreviations: Prod., production; Ser., service.

organizations and the benefits of commonly used management tools through the lenses of improving the productivity and efficiency of organizations. It is evident that production organizations are in the forefront of using management tools that aim to support processes optimization, like benchmarking, outsourcing, and total quality management. Service organizations have a very similar pattern of management tool utilization, with some differences, for example, at the base is the mission and vision statement. A focus on establishing a distinct mission and vision reveals findings about the importance of having a worthy mission in service organizations given the current business environment [63].

At the forefront in Slovenian production organizations are management tools that enable organizations to identify their core processes since outsourcing, coupled with benchmarking, is intensively used. All the preceding ideas reflect the current situation in general, and especially in production organizations in Slovenia, where the key tendency in the last decade has been to focus on core business or core processes, outsource non-value-adding activity, and create a competitive advantage via benchmarking and identifying core competencies [2, 64–69]. The average use of management tools reveals the state of organizational transformation, which is ongoing in production organizations [38, 53, 86] since first and foremost are tools that support organizational optimization. Owing to the importance of optimization tools, those supporting customer relationships have not yet come to the fore [87, 88] and are less frequently used to support work in organizations. On the other hand, in well-developed economies, the most pervasive tools are those designed to strengthen customer relationships, reflecting the organization's orientation toward customers, which is a key concern of organizations in the contemporary business environment [27, 89]. Tools for business process optimization moved to the background, where they still play a role in continuous improvements. From the perspective of service organizations, the pattern of management tool utilization is much more similar to Slovenian production organizations than to the pattern of management tool utilization in well-developed areas, where the most widely used tools are those for enhancing customer relationship management. What is more, it is also evident that in Slovenian service organizations, the level of management tool utilization is lower than in organizations, which again, leaves a lot of space for improvements, with the help of management tool utilization.

In terms of commonly used management tools by organizational department, some results were expected and some surprising. One surprising result was the higher utilization of mission and vision statements in departments where other tools might be preferred, like in fundamental process – Six Sigma, lean concept, reengineering, and total quality management [70–74]. The greater importance of mission and vision statements in fundamental processes certainly does not influence the lower utilization of other typical tools, but it is an indicator that organizations want all their employees to better understand the main organizational directions. Emphasizing the importance of mission and vision statements among employees in organizations has been an important task in the last decade in Slovenian organizations. In terms of expected results, in accounting and marketing, the most used tool is outsourcing. This reflects the current situation in organizations since a large proportion of organizations are small and do not have their own accounting and marketing departments but buy those service on the open market. The high utilization of benchmarking by upper managers indicates a need for comparison with other organizations and establishing best practices in order to improve the operations and performance of organizations.

Looking more closely, tools for supporting competitive comparisons (e.g., benchmarking), building core competencies, and outsourcing [32, 47, 52–54] are widely used by both production and service organizations. In terms of typical management tools aimed at enhancing productivity or the overall efficiency of processes in production organizations, like total quality management, Six Sigma, and lean operations [32, 54, 55], those tools are more commonly used in production than in service organizations, but their level of utilization is low. An exception is total quality management, while Six Sigma and lean operations are last in terms of utilization in organizations.

Regarding the differences in the utilization of management tools in departments in production and service organizations, we see that the utilization of knowledge management in production processes of production organizations is high, which indicates a strong emphasis on knowledge capture and dissemination in organizations in order to retain the knowledge in the organization and to support circulation of knowledge among employees in organizations [57, 75–78].

In terms of service organizations, mission and vision statements dominate, where again the focus is on making employees familiar with the mission and vision of the organization. For instance, a high level of utilization of customer relationship management by top management in organizations reflects the fact that the majority of organizations in Slovenia, as well as in our sample, are small, where the division between operational tasks and management tasks is not very strict. Business practice also supports this since managers in smaller organizations play a key role in customer care and customer relations management [2, 30].

In terms of organizational improvements, as a consequence of management tool utilization in organizations, the improvement of planning activities is the most frequently emphasized benefit in production and service organizations, in various departments. There may be several reasons for perceived improvements in planning by employees, for example, management tools enable recordkeeping, highlight historical trends and decisions, provide simulation techniques within tools for planning, and enable the development of different scenarios.

In terms of linking the performance/output/productivity or efficiency of processes and activities in organizations, the literature offers little comprehensive evidence about the utilization of different known management tools for optimization [5, 55, 56, 79], while reports about concrete results are very rare [62, 73, 80]. Production organizations commonly use tools in lean production, total quality management, or Six Sigma, but there are few practical researches that confirm the positive impact of using different management solutions in terms of production and efficiency improvements [72, 81]. For instance, a study that examined the impact of total quality management, with the mediating role of knowledge management, on business results showed that a combination of total quality management and knowledge management formed a cycle of improvement and development that led to organizational excellence [80]. Next, Six Sigma is a process-focused, statistically based approach to business, and business process improvement that has focused primarily on improving the performance of manufacturing processes. Companies such as Motorola, Allied-Signal, and General Electric have used these tools to produce millions of dollars in bottom-line improvements [62, 82]. On the other hand, the results on the use of different management tools are often hidden in organizations and are treated as a trade secret. Industrial organizations commonly measure the effect of different tools in terms of their productivity or success in terms of financial measures, which is for internal organizational reporting purposes.

Based on our experience with business tools and referring to the several successful examples of using management tools to support process optimization, and along with its organizational productivity, it is evident that a key management tool is often lean manufacturing. This enables production organizations to improve productivity, more closely monitor all process activities, and keep records and logs about production processes. In terms of results, production organizations point most commonly to increases in productivity, shortened production time, evidence about orders, and the traceability of activities.

3.6 Practical implications

In terms of practical implications, our findings could be useful in various ways. Nowadays, newer management tools and those among service organizations' most commonly used management tools have supersede traditional tools aimed at supporting productivity and efficiency, especially in production organizations. In the process of striving for intensive utilization of management tools to support the management of customers and suppliers, production organizations often find themselves in a situation where their focus is much more on efforts to sell their products instead of also on their core business – production. It is important to realize that both production and selling activities are crucial for today's industrial organizations. Thus, an organization needs to have good products and know how to sell them. Other combinations are seldom successful or successful for only a short period of time.

For instance, lean operations and Six Sigma are traditionally associated with improvements in production organizations [62, 72]. In our case those tools are underused, and actions should be taken to intensively use them and take advantage of already proven management tools like lean production and Six Sigma for improving productivity and efficiency in production and service departments in organizations. Thus, the appropriate utilization of lean production could bring substantial benefits for companies since currently production organizations in Slovenia do not rely much on the two mentioned classic management tools, which significantly improve business processes, especially production processes. From the viewpoint of another traditional tool used in production organizations, total quality management, in order to increase the level of quality of processes in organization, it is important to strive for structured strategy with a view to furthering continuous improvement initiatives. For instance, Soare [83] suggests that an efficient and effective approach to business process management and optimization of processes should not be shortsighted or aimed at merely capturing elements from total quality management or reengineering. Rather the approach should reflect a holistic, balanced attitude toward continuous improvement, incorporating, for instance, elements from lean and Six Sigma, not just with regard to tools and methodology but also the level of employee engagement or leadership involvement.

A commonsense and especially balanced approach is needed in organizations about (re)thinking the utilization of their current management tools to support organizational processes. Thus, ad hoc adoption of "typical management solutions" for production core process optimization (like total quality management, Six Sigma, and lean production) and abandoning current widely used tools will not contribute to process optimization. Perhaps there can be some insight in optimization, but other problems can emerge, like losing business opportunities, owing to the shifting exclusive focus on new tools. These results can also cause organizations to rethink their utilization of management tools and carefully set priorities in order to fulfill customer needs for responsiveness and quality simultaneously [5, 84].

In terms of typical management tools, which are used primarily to support production processes, we can offer some practical advice for the utilization of management tools in service organizations. For instance, Six Sigma tools have been focused primarily on improving the performance of manufacturing processes. For instance, Bisgaard et al. [62] in his research emphasized that Six Sigma principles can also be used for the improvement of non-manufacturing, administrative, and service functions. Thus, they argue that the service industry in a modem society represents well over two-thirds of the entire economy and can bring substantial benefits. In this chapter we discuss, using presentations and case studies, how Six Sigma can be used to improve non-manufacturing, administrative, and service processes.

In terms of possible benefits of management tool utilization, users cite improvements in planning activities as a key benefit of management tool utilization in production and service organizations. These findings reveal that the focus is on gaining benefits in planning, which implies that several other areas are still ripe for improvements since management tools could bring about significant improvements in different areas, depending on the purpose and nature of the tool.

Also, our examination of the impact of management tools on improvements in various areas can serve as an important motivator for further and even more intensive utilization of management tools in the future. With the promise that the utilization of management tools will improve productivity and efficiency in all processes in organizations, organizations will show more interest either in investing in management tools or just boosting the utilization of tools that are already in the organization.

To sum up, our results provide an important starting point for becoming familiar with the current state of management tool utilization in production and service organizations, explain how management tools are used within the framework of organizational departments, and provide evidence on the impact of management tool utilization on productivity or efficiency.

3.7 Limitations and future research direction

In terms of limitations, we can emphasize the following ones. First, the sample size we used was small, but it was still enough to draw relevant conclusions. Second, a limited number of management tools was included in the survey. Some organizations may use other management tools not included in the survey. Third, the survey was conducted among Slovenian employees/organizations. Since Slovenia can be seen as a developing economy in Central and Eastern Europe, future generalization of findings to other economies that have different historical development paths could be limited. Finally, the link between productivity/efficiency and the utilization of management tools was assessed by those employees who use the tools. Thus, the possible effect of management tool utilization on productivity was only assessed based on the opinions of respondents.

The most probable directions of future research will be the following. First, expand the current survey to other countries in the region whose historical developmental paths are different than that of Slovenia (e.g., Poland, Hungary, and Austria). In that framework, we could identify the utilization of management tools in organizations operating in well-developed economies as well as in those in a transitional state. Second, to enhance the relevance of this research about the utilization of management tools in various types of organizations, we should check whether organizations use any management tools for improving their work other than those included in our list. Furthermore, new or recently developed management tools should be identified and included in further research. Third, a more detailed survey instrument to identify the link between management tool utilization and their benefits is needed. The most important future research direction will probably be to empirically examine the link between the utilization of management tools and actual productivity in organizations. One way to examine this link would be to ask respondents what productivity was like before and after they started using certain management tools. Despite the possible effect of self-assessment evaluation [85], we would obtain insight into the contribution of tools to increasing productivity; we would also understand which tools contributed most significantly to increasing productivity. A deeper examination of the contribution of single management tools to improvements in various areas in organizations will more comprehensively explain the contribution of those tools to making improvements. Finally, the discussion of the possible benefits of management tool usage could be expanded.

References

[1] Collins J, Porras JI (2002). Built to Last: Successful Habits of Visionary Companies New York, NY: Harper Collins Publishers.

[2] Nedelko Z, Potocan V (2013). The role of management innovativeness in modern organizations. Journal of Enterprising Communities 7: 36–49.

[3] Skarzynski P, Gibson R (2008). Innovation to the Core Boston, MA: Harvard Business Press.

[4] Dumas M, La Rosa M, Mendling J, Reijers HA (2013). Fundamentals of Business Process Management, Berlin: Springer Verlag.

[5] Sinur J, Odell J, Fingar P (2013). Business Process Management: The Next Wave Tampa: Meghan-Kiffer Press.

[6] Slavicek V (2011). Enhancing business process management with knowledge. E & M Ekonomie a Management 14: 123–134.
[7] Magretta J (1999). Managing in the New Economy Boston: Harvard Business School.
[8] Porter M (1998). On Competition Boston, MA: Harvard business school publishing.
[9] Quinn JB (1992). Intelligent Enterprise: A Knowledge and Service Based Paradigm, United States of America New York: The Free Press.
[10] Steffen A (2006). Worldchanging: A User's Guide for the 21st Century New York: Harry N. Abrams.
[11] Drucker PF (1954). The Practice of Management New York, NY: Harper & Row.
[12] Daft RL (2000). Management Fort Worth, TX: The Dryden Press.
[13] Mulej M, Espejo R, Jackson M, Kajzer Š, Mingers J, Mlakar P (2000). Dialektična in Druge Mehkosistemske Teorije Maribor: EPF.
[14] Wren D (2004). The History of Management Thought New York, NY: Wiley.
[15] Dictionary W. (1928). An American Dictionary of the English Language. Available: http://www.cbtministries.org/resources/webster1828.htm. Accessed 2011-09-16
[16] Funk W (1992). Word Origins: An Exploration and History of Words and Language Gravesend: Bell Publishing Company.
[17] Gove P (ed) (2002). Webster's Third New International Dictionary Springfield: Merriam-Webster.
[18] Armstrong M (2006). A Handbook of Management Techniques: A Comprehensive Guide to Achieving Managerial Excellence and Improved Decision Making London: Kogan Page.
[19] Cole G (2004). Management Theory and Practice London: Thomson.
[20] Cooper CL, Argyris C (1998). The Concise Blackwell Encyclopedia of Management Malden: Blackwell.
[21] Linstead S, Fulop L, Lilley S (2009). Management and Organisation: A Critical Text London: Macmillan Education UK.
[22] Lock D (1992). Handbook of Management Aldershot: GOWE.
[23] Morgan J, Wang R (2010). Tournaments for Ideas. California Management Review 52: 77–97.
[24] Pascale R (1990). Managing on the Edge New York: Simon & Shuster.
[25] Potocan V, Nedelko Z, Mulej M (2012). Influence of organizational factors on management tools usage in Slovenian organizations. Inzinerine Ekonomika-Engineering Economics 23: 291–300.
[26] Potocan V, Dabić M (2012). Management Tools in Business Environment Maribor: Faculty of Economics and Business.
[27] Rigby DK (2011). Management tools 2011: An executive's guide. Available: http://www.bain.com, accessed 2014-05-05.
[28] Rue LW, Byars LL (1992). Management: Skills and Application Boston: Irwin.
[29] Van Assen M, Van den Berg G, Pietersma P (2009). Key Management Models: The 60+models Every Manager Needs to Know Harlow: Prentice Hall.
[30] Dabic M, Potocan V, Nedelko Z, Morgan TR (2013). Exploring the use of 25 leading business practices in transitioning market supply chains. International Journal of Physical Distribution & Logistics Management 43: 833–851.
[31] Ooi KB, Lin B, Teh P-L, Chong AY-L (2012). Does TQM support innovation performance in Malaysia's manufacturing industry?. Journal of Business Economics and Management 13: 366–393.
[32] Quinn JB, Hilmer FG (1994). Strategic outsourcing. Sloan Management Review 35: 43–55.
[33] Tamosiuniene R, Jasilioniene R (2007). Customer relationship management as business strategy appliance: Theoretical and practical dimensions. Journal of Business Economics and Management 8: 69–78.
[34] Rigby D (2001). Management tools and techniques: A survey. California Management Review 43: 139–160.
[35] Rigby DK, Bilodeau B (2009). Management tools and trends 2009. Available: http://www.bain.com, accessed 2014-05-05.

[36] Kiezun W (1991). Management in Socialist Countries: USSR and Central Europe New York: Walter DeGruyter.
[37] Newman KL (2000). Organizational transformation during institutional upheaval. Academy of Management Review 25: 602–619.
[38] Sarotar-Zizek S, Milfelner B, Cancer V (2013). Measurement of employees subjective well-being as an aim of social responsibility. Systemic Practice and Action Research 26: 549–560.
[39] Long CS, Abdu Azizl MH, Kowang TO, Ismail WKW (2015). Impact of TQM practices on innovation performance among manufacturing companies in Malaysia. South African Journal of Industrial Engineering 26: 75–85.
[40] Siegel J, Shim J (2000). Accounting Handbook New York: Barron.
[41] Britannica E (2007). Britannica Concise Encyclopedia London: Encyclopedia Britannica.
[42] Buchanan DA, Huczynski AA (2010). Organizational Behaviour Harlow Essex: FT Prentice Hall.
[43] Etzioni A (1997). Modern Organizations New York: Prentice-Hall.
[44] Schermerhorn JR (2008). Management New Jersey, NJ: John Wiley & Sons.
[45] Chatman JA, Jehn KA (1994). Assessing the relationship between industry characteristics and organizational culture: How different can you be. The Academy of Management Journal 37: 522–553.
[46] Maresova P (2010). Knowledge management in Czech enterprises. E & M Ekonomie a Management 13: 131–144.
[47] Porter ME (1996). What is strategy?. Harvard Business Review 74: 61–78.
[48] Kim SY, Jung TS, Suh EH, Hwang HS (2006). Customer segmentation and strategy development based on customer lifetime value: A case study. Expert Systems with Applications 31: 101–107.
[49] Reichheld FF (1993). Loyalty-based management. Harvard Business Review 71: 64–73.
[50] Kaplan RS, Norton DP (2008). Mastering the management system. Harvard Business Review 86: 62–77.
[51] Mintzberg H (1994). The fall and rise of strategic-planning. Harvard Business Review 72: 107–114.
[52] Cooper RG, Kleinschmidt EJ (1995). Benchmarking the firms critical success factors in new product development. Journal of Product Innovation Management 12: 374–391.
[53] Jerman D, Zavrsnik B (2012). The model of marketing communications effectiveness: Empirical evidence from Slovenian business-to-business practice. Journal of Business Economics and Management 13: 705–723.
[54] Harrison B, Kelley MR (1993). Outsourcing and the search for flexibility. Work Employment and Society 7: 213–235.
[55] Powell TC (1995). Total quality management as competitive advantage – A review and empirical-study. Strategic Management Journal 16: 15–37.
[56] Davenport TH, Short JE (1990). The new industrial-engineering – Information technology and business process redesign. Sloan Management Review 31: 11–27.
[57] Bures V (2009). Conceptual perspective of knowledge management. E & M Ekonomie a Management 12: 84–96.
[58] Farndale E, Paauwe J, Hoeksema L (2009). In-sourcing HR: Shared service centres in the Netherlands. International Journal of Human Resource Management 20: 544–561.
[59] Pal SK, Kapur V (2009). The Gradual Adoption & Maturity of Corporate Blogs Norristown: Int Business Information Management Assoc-Ibima.
[60] Meidute I, Davidaviciene V, Raudeliuniene J (2012). Research on Applying Radio Frequency Identification Technology at Lithuanian Companies. In: 7th International Scientific Conference Business and Management 2012, Ginevicius R, Rutkauskas AV, Stankeviciene J (eds) Vilnius, Lithuania: Vilnius Gediminas Technical Univ Press, Technika 989–996.
[61] Malefyt TD (2009). Understanding the Rise of Consumer Ethnography: Branding Technomethodologies in the New Economy. American Anthropologist 111: 201–210.

[62] Bisgaard S, Hoerl RW, Snee RD. (2002). Improving business processes with six sigma. Quality Congress. ASQ's Annual Quality Congress Proceedings, 701–704.
[63] Malbašić I, Rey C, Potočan V (2014). Balanced organizational values: from theory to practice. Journal of Business Ethics 130(2):437–446.
[64] Hauptman L, Horvat M, Korez-Vide R (2014). Improving tax administration's services as a factor of tax compilance: The case of tax audit. Lex Localis-Journal of Local Self-Government 12: 481–501.
[65] Potocan V, Mulej M (2009). How to improve innovativeness of small and medium enterprises. Management 14: 1–20.
[66] Potocan V, Nedelko Z (2014). Management innovativeness: A case of Slovenian small and medium enterprises. Transformations in Business & Economics 13: 41–59.
[67] Potočan V, Mulej M (2007). Transition into Innovative Enterprise Maribor, Slovenia: Faculty of Economics and Business.
[68] Sarotar-Zizek S, Milfelner B, Mulej M, Breg T, Potocnik A, Hrast A (2012) "Empirical data about social responsibility in Slovenia", presented at the Idimt-2012: ICT Support for Complex Systems, Jindřichův Hradec, Czech Republic, September 12–14 2012. Publisher: Johannes Kepler Universität, Linz, 215–229.
[69] Zavrsnik B, Potocan V (2012). Development of the Slovenian textile and clothing industry. Industria Textila 63: 204–207.
[70] Eckes G (2001). The Six Sigma Revolution: How General Electric and Others Turned Process into Profits New York: John Wiley & Sons.
[71] Mikel H, Schroeder R (1999). Six Sigma: The Breakthrough Management Strategy Revolutionizing the World's Top Corporations New York: Doubleday.
[72] Pakdil F, Leonard KM (2014). Criteria for a lean organisation: Development of a lean assessment tool. International Journal of Production Research 52: 4587–4607.
[73] Stefanic N, Gjeldim N, Mikac T (2010). LEAN concept application in production business. Tehnicki Vjesnik-Technical Gazette 17: 353–356.
[74] Womack J, Jones D, Roos D (1991). The Machine that Changed the World: The Story of Lean Production New York: Harper Business.
[75] Becerra-Fernandez I, Gonzalez A, Sabherwal R (2004). Knowledge Management: Challenges, Solutions, and Technologies New Jersey, NJ: Pearson-Prentice Hall.
[76] Becerra-Fernandez I, Sabherwal R (2010). Knowledge Management Systems and Processes New York, NY: M. E. Sharpe.
[77] Easterby-Smith M, Lyles MA (2672). (2006, 24 February 2016). Handbook of Organizational Learning and Knowledge Management. Available: http://www.blackwellreference.com/public/book.html?id=g9780631226727_978063127.
[78] Frost A (2010). Knowledge Management Best Practices. Available: http://www.knowledge-management-tools.net/KM-best-practices.html.
[79] Sahay BS, Gupta JND, Mohan R (2006). Managing supply chains for competitiveness: The Indian scenario. Supply Chain Management-an International Journal 11: 15–24.
[80] Daud S (2009). The Influence of Total Quality Management and Knowledge Management on Business Results Kuala Lumpur: Business Information Management Association.
[81] Dodangeh J, Rosnah MY, Ismail N, Ismail YM, Biekzadeh MR, Jassbi J (2012). A review on major business excellence frameworks. Technics Technologies Education Management-TTEM 7: 1386–1393.
[82] Kanji GK (2008). Reality check of six sigma for business excellence. Total Quality Management & Business Excellence 19: 575–582.
[83] Soare P (2012). Opportunities for Driving Continuous Improvement through TQM, LEAN and Six Sigma within Business Process Management. Proceedings of the 6th International Management Conference: Approaches in Organisational Management. Faculty of Management, Academy of Economic Studies, Bucharest, Romania, vol. 6(1), pages 193–202, November.

[84] Chopra S, Meindl P (2013). Supply Chain Management: Strategy, Planning and Operation Harlow, United Kingdom: Pearson.

[85] Schwartz N (1999). Self-reports: How the questions shape the answers. American Psychologist 54: 93–105.

[86] Nedelko Z, Potočan V (2016). “Management practices utilization in organizations – A comparison between catching up and well-developed economies,” Management (Croatia) 21: 1–20.

[87] Telnova AV, Bezbozhnyi VL (2011). “Improving the process of strategic planning at an enterprise,” Actual Problems of Economics:235–242.

[88] Zagorniak NB (2012). ”Consumer value within business unit management system: A compass in new reality and the battle for market opportunities”, Actual Problems of Economics:164–171.

[89] Phillips D (2006). “Relationships are the core value for organisations,” Corporate Communications: An International Journal 11: 34–42.

Teresa Carla Oliveira and Stuart Holland

4 Economic and social efficiency: the case for inverting the principle of productivity in public services

Abstract: Productivity has been the benchmark of economic efficiency. Through better organization and innovation, including the use of new technologies, enterprises in an efficient economy produce more with either the same or less labor so that output per employee rises. Productivity criteria also have been typical of the New Public Management introduced in the UK and the USA since the 1970s. Yet in the sphere of production, higher productivity may outcome in structural unemployment in society as a whole. In the social domain, such as seeking more output per health worker, social worker, or teacher, it also may lower the quality of service. This chapter distinguishes economic from social efficiency. It submits that an efficient society will invert the principle of economic productivity by employing more people in education, health, and public and social services with smaller class sizes, shorter waiting lists for hospital admissions and more personalized services for those in need, not least in societies with aging populations. Drawing on evidence from the USA, Europe, and Japan, it suggests that this has implications for redressing technological unemployment and increasing welfare within a social and economic paradigm of lean production.

> Conventional notions of productivity are pretty hopeless when it comes to describing what doctors, teachers and social workers actually do. Consider a classroom with a sensible number of pupils. Want to increase that teacher's productivity? Add 30% more kids. What about those hospitals that take so long to get patients well and return them to the community? Speed it up, for goodness sake! Get them out of bed and back on the streets if you want to hit that productivity target – in the UK's case, that of waiting times. And, while you are at it, fix revolving doors to the front of the building so they can get back in quicker when they fall ill again.
>
> **Stefan Stern, It is time to end our unproductive fixation with productivity. The Financial Times, 11 April 2006**

> His arguments that gains from flexible production should be shared both between management and labour and within society is of the first importance to who benefits in what way from both an efficient economy and an efficient society.
>
> **Jacques Delors, Foreword to Stuart Holland, *The European Imperative: Economic and Social Cohesion in the 1990s*, 1993**

https://doi.org/10.1515/9783112206775-004

4.1 Introduction

Economic productivity has been the benchmark of efficiency not only since Adam Smith illustrated the division of labor in pin making in his *Wealth of Nations* (1776) but especially since Henry Ford combined mass production with a Taylorist ultra division of labor [1]. An efficient economy produces more with either the same or less labor so that productivity increases. By contrast, however, as proposed in this chapter, social efficiency in the sense of more labor-intensive and personalized public services inverts economic productivity. No one ranks a primary or secondary school or university as better than another because it has bigger classes and less personalized teaching. No one prefers a hospital or health clinic because its doctors or health workers treat more patients than another. An efficient society will employ more people in education, health, and public and social services with smaller class sizes in teaching, shorter waiting lists for hospitals, and more personalized and swifter social services for those in need, not least in societies with aging populations.

It has tended to be assumed that social welfare can be afforded only if an economy first becomes more competitive, that is, that competitiveness must precede welfare. Yet the implicit logic of this is flawed in that it fails to recognize the circularity of expenditure and income. More health workers, teachers, and carers spend more than fewer. As their jobs are created, they both raise the level of employment and increase effective demand. They pay not only indirect taxes from their expenditures but also direct taxes on their income since public services are in the "overground" economy [2].

Milton Friedman [3] asserted that markets are always more efficient than governments. This has been challenged by the speculation in toxic financial derivatives that gave rise to the subprime crisis and the greatest crash of markets in Western economies since 1929. It also has been challenged not only by how state intervention resolved the 1930s Great Depression, as with the US New Deal [2], but also by evidence that the state not only can provide the conditions for markets to flourish, but also create markets directly. As with computers, since Alan Turing did in breaking the German Enigma codes in World War II [4], their development thereafter by US Defense Department funding, the World Wide Web, as well as US federally sponsored innovations such as the algorithm of Google and the creation of nanotechnology [5].

Friedman [3] also claimed that public investment and spending "crowds out" and drains the private sector. Yet it has been overlooked that he also claimed that this would be true only in the case of full employment, whereas unemployment is now more typical of many advanced economies, while public investment generates demand for private sector goods and services. For example, investment projects financed by the European Investment Bank generate up to three times their initial investment, employment, and income in the private sector. Thus, if European Investment Bank renovates a hospital, as they have done 365 times in France, the renovation is done by private construction companies, which engage other private sector companies to carry out the project [2].

It has been widely assumed that economic efficiency *precedes* social welfare and that growth is a necessary condition for redistribution, which is a misconception. Economic efficiency historically has been derived *from* rather than *preceded* social efficiency. It depended on not only natural talents and inspiration of the kind that Schumpeter [6] submitted were characteristic of innovators, but also the state assuring an educated, literate, and numerate general workforce [7].

These are social external economies for which firms may pay through taxation – if not avoiding them through transfer pricing [2] – but cannot directly provide for the population as a whole. There are also other social external economies. Shorter waiting times for more frequent public transport, shorter waiting lists in hospitals, and a less polluting environment all improve economic efficiency in terms of less time to get to work, or being healthier in getting there, or obtaining treatment. Fifty years ago Edward Denison's study for the Brookings Institution [8] *Why Growth Rates Differ* found education and health to be among the most consistent factors explaining why higher welfare countries have better economic performance. In these key senses, increased economic efficiency *proceeds from* rather than precedes from social welfare and well-being.

In addition, income redistribution also sustains rather than drains economies. In a recent study for the European Central Bank, Carroll et al. [9] found that a tax transfer to the lower half (in terms of net worth) of the population or the unemployed is two to three times more effective in increasing aggregate spending than a stimulus of the same size for higher earners. Moreover, those dependent on state pensions have a high propensity to consume and sustain demand while the richer tend mainly to save. Cutting state pensions does not improve competitiveness since few people eligible for pensions also work for companies. Cutting pensions on the grounds that there is a fiscal crisis, such as in the EU, ignores the fact that this was caused by speculation by banks and hedge funds, which both challenges the principle that markets are necessarily efficient and defies both the principles of social justice and conviction among people that governments can govern rather than markets and market values rule.

A higher level of public expenditure in turn is crucial for lowering unit costs in the private sector, where economies of scale and more of the same, despite economies of scope and more with the same, remain important. The lack of internal EU demand as a factor constraining competitiveness has been stressed time and again in the debate on the Eurozone crisis and the limits of quantitative easing [2, 10, 11], that is, that low or zero interest rates will not by themselves promote a sustained recovery, not least since, as Keynes stressed, when confidence in a recovery is low, expectations of returns on net investment are depressed [12, chapter 12] and low interest rates are like pushing on a string rather than recovering new investment.

In terms of economic efficiency and productivity, it is also widely presumed that employing more people in public services is only a cost rather than a gain from investments in health, education, and social wellbeing. Certainly more labor-intensive employment is not a sufficient condition for a more efficient society. This also depends on the quality of more personalized services and to efficiency in outcomes. But, as illustrated in more detail later, more time to individualize a service, such as health care, raises its quality for both the public and service providers, whereas higher productivity at the cost of worker overload and burnout does not.

A Fordist and Taylorist logic has been reflected in the concern of New Public Management (NPM), since the 1970s, to increase efficiency in social domains such as health and education. Within NPM the major concern has been to assure a high-performance work system (HPWS) by imposing market or "quasi-market" performance criteria at operational levels [13], often using a transactional approach through financial incentives and managing people within a human capital framework of seeking to raise the rate of return on employees [14]. By contrast, social efficiency implies a service within a high-involvement work system (HIWS) through a more transformational leadership approach for which relational models [15] of management are more appropriate, within a longer standing human relations perspective.

It is on such grounds that this chapter proposes that social productivity should be the inverse of economic productivity, assuring more teachers and smaller classes, more health workers and shorter waiting lists, and more carers for the aged. It cites endorsements of this by European heads of state and governments intended to relaunch and reinforce a European social model, which have been displaced by the contrary demands for "structural reforms" aiming to reduce employee rights since the onset of the Eurozone crisis. It illustrates that some of the highest productivity gains in the world, by leading Japanese companies, have been by assuring core employees lifetime employment and that, if they suggest continuous improvements in work methods, their productivity gains are to the mutual advantage of themselves and their work colleagues as well as their organizations.

The chapter critiques the idea that efficiency and innovation can only be achieved by reducing employee rights through more flexible labor markets by which outsiders can take insiders' jobs at lower cost, which has been the *leitmotif* of the "structural reforms" demanded by the troika of the European Commission, the European Central Bank, and the International Monetary Fund (IMF) since the onset of the Eurozone crisis in 2009 [16]. It illustrates that there is no basis for this in the purely theoretical "insider-outsider" thesis [17] of former Nobel Economic Committee member Assar Lindbeck and the British economist Denis Snower, which was highly influential in persuading German employers from 2004 to demand longer hours for no increase in pay, which reduced the growth of internal demand in Germany to the detriment of the exports of other EU member states with a "beggar-my-neighbor" rather than "better-my-neighbour" syndrome.

The chapter also cites recent recognition from the IMF that there also is no basis for "structural reforms" enabling more flexible labor markets and substantiates the claim that leading Japanese firms have been able to achieve some of the highest productivity rates in the world through flexible and lean production by offering core workers lifetime employment and profit sharing. It shows that it was the failure of flexible labor markets in Japan in the early twentieth century that gave rise to such high productivity, as well as to annual wage increases for core workers, and how these enabled the *kaizen* of continuous improvement in methods of work operation and thus sustained competitive advantage.

It develops the concepts of both economic and social efficiency in relation to theories of transformational and transactional leadership and shows that case studies of leader-member exchange (LMX) in the West have shown the importance of retaining core workers to achieve sustained productivity [18, 19, 108]. It recognizes the degree to which theories of institutional logics such as those of Lok [20] have shown that government reforms can be disregarded at organizational levels and suggests the complementary distinction of organizational and operational logics [21] in relation to the role of operational learning, which is excluded in a Fordist Weberian paradigm, which insists on top-down performance appraisal. The chapter maintains that hierarchy and surveillance in education, as claimed by Foucault [7], inhibit creativity and efficiency in NPM in both education and health services and that, in seeking to introduce market criteria, NPM has been counterproductive in the UK [22, 23].

By contrast, the chapter illustrates from other international examples how efficiency in health care provision can be achieved by learning up from lean and flexible production within a post-Fordist and post-Weberian paradigm. In this sense it draws on relational models (e.g., [15]) of management by involving employees in decision-making and gaining commitment to high performance through concern with the well-being of both providers and users of services. It then shows how such arguments were endorsed by successive European Councils before the onset of the Eurozone crisis, thereafter were displaced and denied, but could and should in principle be recovered to achieve the economic and social efficiency that the European Union needs for its own democratic legitimation.

4.2 Productivity and economic efficiency

One of the claims made by Milton Friedman [3], and which influenced Margaret Thatcher – through briefings from a fellow member of parliament and close confidant, Sir Keith Joseph – is that markets always were more efficient than governments. It encouraged her to endorse a presciently named “Big Bang” liberalization of financial markets in the UK in the mid-1980s. This was part of the so-called supply side economics that posed a challenge to John Maynard Keynes’s [12] notion that the state should intervene to manage the level of demand since markets could not of themselves assure a balance of demand and supply. This was paralleled from the 1980s by theories of efficient markets on the assumption that, provided the markets were not constrained by government regulation, perfectly informed entrepreneurs with "rational expectations" would ensure an optimal allocation of resources [2]. Influenced by such a theory, US Treasury Secretary Larry Summers persuaded the US Congress in 1999 to repeal the Roosevelt New Deal separation of deposit banking from speculative finance. The outcome, within a decade, was the worst financial crisis to hit the Western world since the Crash of 1929. This then generated a crisis in the European single currency area, the Eurozone, from which it has not yet recovered [2].

4.2.1 Flexible labor markets

The supply side case for the alleged superiority of efficient markets rather than government regulation of finance was paralleled by the case that governments should also deregulate labor markets by structural reforms [16]. Their assumption was that if companies could fire workers more easily, then new and innovative firms would be able to attract them to expand their businesses. Likewise, the lower labor costs that the new enterprises would incur from the reduction of social protection would increase the competitiveness of Europe as a whole in the face of globalization. This narrative has been extremely well marketed by the European Commission and related institutions and interest groups since the onset of the Eurozone crisis [24]. Yet the narrative, like many others from neoliberal economics, such as that macroeconomic austerity is the only way to resolve high unemployment and low growth, is a myth.

For example, a work by the IMF's research staff in its April 2016 *World Economic Outlook* recognized that while productivity can be increased by innovation, through investing more in research and development, training, and using more highly skilled, there is no evidence whatsoever from any of the advanced economies of the Organization for Economic Cooperation and Development (OECD) of negative effects on total productivity of social protection for workers. Because labor market deregulation has been a key ingredient in the structural reforms and structural adjustment austerity programs demanded by the troika of the IMF, the European Central Bank, and the European Commission in several European member states, this represents a direct challenge to the intellectual and ideological basis of how the crisis since 2009 in the Eurozone has been mismanaged [24].

The case for flexible labor markets was made most notably in 1988 by Nobel economic committee member Assar Lindbeck and British economist Dennis Snower (1988), who proposed that Europe could be competitive only by reducing the right of so-called insider employees to defend high wages and benefits against outsiders, who would be willing to work for less. The theory was very influential with leading German employers, who, led by Siemens, from 2004 demanded that employees work an additional 4–5 h a week, without any increase in pay, in order to raise productivity. This demand was met on the undisguised threat that, if workers did not comply, they would relocate to former Soviet satellite states in Central and Eastern Europe or to China – which they did anyway [2].

By contrast with such top-down hierarchical presumptions of how to gain greater efficiency, the role of operational managers is vital. This can be informed by a distinction within LMX theory between transactional and transformational leadership and how to achieve a work climate by which employees at operational levels who are the front line of the potential for achieving greater efficiency and enhanced customer/user satisfaction can do so.

Moreover, although it was replete with premise-dependent algebra, the Lindbeck-Snower insider-outsider claim [17] offered no evidence, from any country, to support its case [25]. On the contrary, shortly after Lindbeck and Snower came out with their idea and well before more recent IMF findings were published [113], extensive evidence refuting their claim was published by the OECD [26] and by independent researchers [27].

The internal restraint of wage increases in Germany then had beggar-my-neighbor effects on other EU economies by reducing demand for their exports. In addition, by contrast with the demand from 2004 of leading German firms that employees should work an additional 4–5 h a week for no increase in pay to make them more competitive, Germany in 2004 overtook the US as the world's leading exporter and did so with a population less than a third that of the USA, indicating that it already was more than three times as productive in terms of national output per capita as the USA.

On the London launch in 1988 of the Lindbeck-Snower book, one of us asked Dennis Snower how his and Lindbeck's insider-outsider thesis related to Japan, to which he replied, "I don't know about Japan. It may be different." It is, and has been so for a century, precisely because at the peak of the industrial revolution in Japan, which had been inaugurated by the Meiji dynasty in the later nineteenth century, by the opening decades of the twentieth century flexible labor markets failed in the sense that industrial firms trained engineers at considerable cost but then found that they were poached by others offering them higher pay [28].

4.2.2 Flexible and lean production

It was in response to this failure that leading Japanese firms offered core workers lifetime employment to the age of 55, combined with seniority pay – annual wage increases – and biannual profit sharing. It was this policy that enabled Japanese companies to achieve formidable productivity gains through the continuous improvement of *kaizen*. *Kaizen* [29] in Japanese combines two words and two concepts: *kai* refers to improvement and *zen* signifies something that is to mutual advantage, just the inverse of the Western.

This enabled Japanese companies to achieve highly flexible and lean production [30], that is, precisely the inverse of the Western flexible labor market model. Western firms have treated capital as a fixed cost and labor as a variable cost, which can be reduced either by cutting wages or letting people go in a recession. In both large and many medium-sized companies in Japan, this can be the case for temporary workers, but it is not the case for core lifetime workers. It is because employment for such insiders is secure that such Japanese workers have known that if they suggest improvements in methods of work organization that cut waste in terms of materials or time and thereby go "lean," they will not innovate themselves – or their colleagues – out of a job.

In Japan, following the 1973 first OPEC oil shock and encouraged by the Japanese Ministry of International Trade and Industry, *kaizen* went big, registering widespread, macro effects [31]. With financial support, leading companies scrapped much of their earlier capital investment and accelerated continuous improvement in methods of work organization, by which companies such as Toyota and Honda managed to double labor productivity in the following decade, whereas the US auto majors managed next to none [31]. By the early 1990s, Toyota was receiving up to 95 suggestions for continuous improvement per employee per year and was implementing many of them, whereas Detroit was not even soliciting such suggestions from their workers [32]. The direct reward for an accepted proposal in process innovation was not great. But the indirect gain was to the mutual advantage not only of the individual employee but also other employees since the companies survived and thrived and since a high proportion of income was through profit sharing.

Notably, lifetime employment in Japan, like the seniority wage system, is a norm, not a rule. Like a psychological contract [33, 34], it has been implicit in practice rather than explicit in the sense of an employment contract.

Nonetheless, where the Japanese flexible production paradigm has been open to question is in its lack of work-life balance. For decades leading Japanese companies paid limited attention to this, tending to demand that workers commit themselves to their flexible production schedules. This has coincided with the cultural phenomenon in Japan of *karoshi*, or "working to death." According to the Japan Productivity Center (2009) nearly 90% of workers reported that they did not even know what was meant by work-life balance. The Japanese Trade Union Confederation found that two-thirds of men put in more than 20 h of unpaid overtime monthly. One in 25 admitted to working 80 extra hours, a level that risks *karoshi*.

Leading Japanese firms have gone to considerable lengths in seeking to offset the trend toward *karoshi* [36], such as by insisting that employees leave work at 5:00 pm, albeit so far with limited success since many white-collar workers simply take work home with them. What this chapter suggests, however, and was recommended by the European Council to heads of state and governments at Lisbon in 2000, is that the principle of *innovation-by-agreement* derived from Japanese flexible production should be matched in Europe by the right to a work-life balance. However, while it was recommended both to the Lisbon Council [38] that such work-life balance should have been a European citizen's right, the Council endorsed this only as a recommendation, which, arguably, is among the reasons why the aspirations of the Lisbon Agenda as a platform for European innovation and competitiveness failed to gain traction, as did the parallel aspiration of ministers at Lisbon to regenerate and enhance the European Social Model.

4.2.3 Transformational versus transactional change

What Japanese manufacturers learned from the Toyota production system was transformational change. This was implicit in Schumpeter's [6] case that it is product and process innovation rather than only reducing costs that lifts economies and societies to higher levels of both efficiency and wellbeing. In addition, Weber [39] considered entrepreneurial innovation and leadership an example of his three leadership archetypes, with the others being traditional and hierarchical (as in feudal societies) and bureaucratic. Weber also recognized that charismatic leadership could succeed in the short term but fail thereafter if a charismatic leader did not establish effective management structures. Whereas process innovation in the sense of doing something better does not need to be top-down, innovated from on high, but more typically, as in the case of Japanese *kaizen*, base-up, from operational levels. Such innovation also can be transformational, as it has been for Toyota and other leading Japanese companies.

By contrast, transactional leadership tends to be financial such as in "A fair day's work for a fair day's pay" from an entrepreneur or an organization. But this may have to do with offers that workers feel they cannot afford to refuse if unemployment is high. The pay – though it was high for Ford [1] – may not be high, but it is better than the close-to-subsistence poverty-line levels that many workers who are able to find work have had to accept since the financial crisis in the USA, where average incomes have not risen since the early 1970s, or in countries subjected to so-called structural adjustment policies that have led to reductions in labor protection in the Eurozone since 2008.

One of the clearest statements of transactional leadership is leadership-exchange theory [18, 19, 108] in terms of role taking, role making, and rewards for performance within what otherwise remain routine work methods. But, as stressed by Lunenberg [40], LMX theory works best when it assumes two groups. In-group members are given greater responsibilities, more rewards, and more attention. The leader also allows these members some latitude in their roles. They work within the leader's inner circle of communication. By contrast, out-group members are outside the leader's inner circle, receive less attention and fewer rewards, and are managed by formal rules and policies.

Such findings, from western rather than Japanese organizations, directly conflict with the ungrounded Lindbeck-Snower thesis. Parallel findings also indicate that in-group members [19, 41] have higher productivity, job satisfaction, and motivation and engage in more citizenship behaviors than out-group members. George and Jones [42] also submit that in-groups should be as large as out-groups, which in leading companies in Japan has been the case for core workers relative to temporary or part-time workers [28, 37]. In-group members also tend to gain from relational models [15], reflecting high involvement rather than only HPWS.

Thus the initial and sustained improvement was not transactional but transformational, in the sense of transforming a company such as Toyota, which in 1948 had been producing only as many vehicles in a year as General Motors (GM) was producing in a day, into a global giant that in 2006 overtook GM as the world's leading auto company. This is what emerged as a post-Fordist production paradigm [25, 110] in the sense of enabling economies of scope – more with the same – rather than economies of scale – more of the same – even if, from the 1970s, leading Japanese firms were able to combine both.

4.2.4 Operational and organizational logics

Bass [43, 44] defined transformational leadership in terms consistent with LMX theory in the sense of how this may encourage employees to trust, admire, and respect the transformational leader. He identified three features of this: (1) increasing workers' awareness of task importance, (2) persuading them to focus on both team and organizational goals, and (3) recognizing and enabling fulfillment of their own personal aspirations and ambitions. Most institutions locked into a Weberian, Fordist, and Taylorist organizational logic are not centrally concerned with any of these three recommendations [43, 44]. Nor are they predisposed to learning from tacit knowledge or implicitly acquired skills at operational levels because they already have determined what should be both known and done within their own presumptions of what is the "one best way" of management [45, 111].

In practice, most upper-level managers still presume that organizational change will form by itself, its meaning transmitted to divisional, unit or line managers, understood at face value, and acted on. They do not see a need to solicit suggestions from employees, including middle managers, for improvements in operational practice, which could have significance both for the value of the products or services they offer and for the efficiency of the organization as a whole. Nor do they recognize that, by reflective practice with employees [46–48], they could open up new skill paths for individuals and groups, new innovation trajectories for an operational unit, department or division, with wider learning implications for the entire organization.

This relates to questions of efficiency between the "center" and "periphery" of Europe, which have played such a prominent role in debates on how to respond to the

Eurozone crisis, with the presumption of German finance minister Wolfgang Schäuble that the economies of the periphery should compete with Germany by reducing their labor costs [2, 49]. This neglects differences in economic structure, such as in Germany, which is a highly industrialized modern economy, whereas Greece and Portugal are not. It also neglects differences in the size and scope of firms, most of which are traditional, regional, or local in the southern European periphery, with some multinationals, as in Ireland, which has benefited – not least from a common language – in attracting US high tech multinationals.

In addition, rather than assuming that central economies are saints and peripheral economies are sinners in terms of a Protestant work ethic [50], when the nature of firms is the same – as in the parent or the subsidiaries of a multinational company – the periphery may be more efficient than the center. Such is the case with the Portuguese subsidiary of Volkswagen, which has been more efficient on all internal efficiency criteria of Volkswagen than any of its plants in Germany, including its iconic Wolfsburg plant [51].

4.3 Social efficiency

The words efficiency and economy are often juxtaposed. This is less so for efficiency and societies. Yet a society is not efficient if it allows banks to speculate with people's savings and destroy them. Nor is it efficient if it tolerates persistently high levels of unemployment. Nor if a sense of injustice threatens the legitimacy of its institutions. Nor if it only serves markets rather than ensures that they also serve people. Nor if it demands that economic criteria for productivity should obtain without qualification in social domains such as health, education, and public services.

For a productivity logic is counterproductive in the social domain. An efficient company may take labor out as one way to "go lean" and aim for a HPWS or on the basis of people's ability, motivation, and opportunity (AMO) perspective [52, 112]. But an efficient society puts people in and goes lean in other ways, such as in a more flexible use of plant and equipment, cutting waste in time and procedures, and ensuring personalization of student or patient focus in education or health services. In this case, the concern is with achieving HIWSs through relational models of managing people at work that goes beyond the AMO framework by strengthening relationships and mutual advantage between employers and employees [15].

The earlier claim that no one judges a hospital or health clinic with more patients as being better than another is clearly corroborated in the case of nursing by Oppel and Young [53], who, along with others such as Everhart et al. [54], echoing also Spetz et al. [55], and earlier research by Mark, Harless, McCue and Xu [56], found that higher nurse-patient staffing is associated with better clinical patient outcomes such as dealing with complications gaining, lower mortality rates, and higher patient perceptions of the quality of health care.

Nor is it clear that economic efficiency can be achieved in the social sphere only through financial incentives. Thus, in health care, Brosig-Koch et al. [57] found that the outcome of fees-for-service payments rather than capitation (CAP), in the sense of payment for the number of patients treated, resulted in less patient-oriented care than CAP after the choice of payment scheme was made and did not necessarily improve but could even worsen patient treatment. From a large administrative data set in the Netherlands, Douven et al. [58] found that more altruistically than financially motivated providers of mental health care achieve better treatment outcomes. Nor is the mode of ownership versus organization vital to social efficiency. Data from Germany on whether hospitals are public or private show comparable results rather than superior performance by private providers [59].

4.3.1 Learning from flexible production

Stefan Stern's observations, which head this chapter, clearly are ironic. Yet they point to a central dysfunction of paradigms of change management that draw explicitly on performance criteria such as productivity from the commercial sphere without recognizing their illogic in the social sphere. For the implicit logic of an efficient society it would employ more people rather than fewer, both in the sense that doing so is central to economic and social cohesion and precisely because people want smaller classes in teaching and more personalized care in health or social services. The market itself shows this in the degree to which those who can afford it pay for it. Thus, there is a demand for more and better jobs [113] and a human approach to managing people that implies the use of relational models [60].

Not that this in itself means that either education or health or other social services have no room to make efficiency gains. One of the main arguments in several of the papers to which we recently have contributed (e.g., [25, 61–63]) is that paralleling flexible production in the provision of health care services can reduce the underutilization of operating rooms or wards by increasing their flexible use and cut waste in patients' waiting time for treatment. But this is different from the operational logic of British health and education reforms, which have tried in a Fordist manner to produce more for less by increasing pupil/student or patient throughput within an unchanged organizational paradigm.

Failure to address the issue of what exactly an efficient economy and an efficient society also carries external and internal costs in the political sphere. This is evident in the reaction to both globalization and European expansion, not only in Western Europe but also in Central and Eastern Europe, where Kregel et al. [64] correctly forecast that "market shock" and the lack of counterpart policies for social cohesion would cause "withdrawal" to earlier models of the primacy of nationhood, increased xenophobia and more overt racism, thereafter echoed by Boyes [65] and Wagstyl [66] a decade before the onset of the refugee and asylum-seeker crisis to which European governments had no common response.

4.3.2 Deigning down and "deliverology"

Top-down change in the name of reforms designed to increase productivity has been typical of the introduction of NPM in the UK but has been almost entirely counterproductive. For example, Seddon and Donovan [67], drawing on Blond [68], as well as on their own research, have shown how an ideological approach to changing management in the public sector has driven it in the wrong direction. The New Labour governments intensified performance criteria and a culture of "deliverology" [70], which was deigned down by imposition on public sector managers and employees and which Tony Blair has since claimed to be his main political legacy and current global mission [71, 72]. Yet it did not work.

For example, one of the "deliverology" reforms that the New Labour government introduced was on the basis that "back-office" activities in delivering social services could be cut while enhancing "front-line" services. Early in the New Labour government, Gordon Brown, Chancellor of the Exchequer, declared that there would be no investment in public sector services without higher performance. The Department for Work and Pensions persuaded him to invest £200 million in a new system for the delivery of housing benefits by local authorities. It was one of the earliest attempts to impose a separate front-back office design, the front being the means of access and the back being the place where claims are processed. The two were to be linked by document image processing with information received in the front office scanned and sent electronically to the back office.

Taylorist time and motion targets were central for the new change in management design. For the front office these included how quickly people were seen, how long phones rang before being picked up, and how soon documents were scanned and sent to the back office. The back office needed to report on the response time for the processing of correspondence, how many "work activities" were done, and the time it took to process a claim. But, as Seddon and O'Donovan stress, the flaws in the design soon became apparent by something intuitively obvious: someone arriving without all the required documentation would need to establish eligibility and entitlement. If they could not, the delivery stalled.

This led to backlogs in housing benefit offices all over the country. On the advice of the Department for Work and Pensions, local authorities then hired "backlog-busting" services from the private sector, costing tens of millions of pounds. But this ignored "the human factor." People are not walking data processors. Until the new system was introduced, they might rarely have been obliged to produce a birth certificate, a marriage certificate, a divorce certificate or a national insurance or social security number. If they had hit an early or midlife crisis, they might know where in principle some of these were, with a former partner, but not be able to access them. Information was frequently "lost" before it could be processed, and thus people were often asked to bring it in again.

The previous practice in which someone might apply for a benefit without bringing in their national insurance or social security number, and just being asked to bring it in next time, if they could, was abolished. Or, if they could not remember it, a social worker would find it for them and ensure that their claim therefore was duly processed. Understandably, this not only was dysfunctional but provoked anger from people whom the new "improved" system had reduced to numbers and whose rights the system was blocking and denying. As a consequence, social security managers felt compelled to put up posters announcing that any offensiveness to staff would result in legal action by management [67].

There are also pressures resulting from understaffing. An example from a current study directed by one of us is from an emergency unit in a pediatric hospital involving a mother with a baby in her arms waiting to be seen for more than half a day, under stress and voicing her concern because the baby was not allowed either to drink or eat before being seen, to which a nurse replied that she should "not be concerned because the symptoms of the [as yet unevaluated] baby are alright." Of course, especially in public hospitals, with pressure from governments to gain more patient throughput per staff member, there is less and less time to attend to individual patients. The same goes in education, with a ranking of schools by the number of examination passes, where there is less and less time to individualize learning.

4.3.3 The limits of performance appraisal

There have been parallel failures in the demands of NPM to impose greater accountability of public sector employees through intensified performance appraisal, in particular, by new line managers brought in from the private sector. Organizational psychologists such as Fletcher [73] have recognized that the criteria for performance appraisal not only are fraught in practice but also in principle.

Thus Fletcher recognizes not only that personnel assessment and performance appraisal actually may have no benefit for the person appraised but also may offer no gain for an operational unit or an organization if the appraisal and evaluation are not to mutual advantage. For example, employees should be able to indicate not only what may be going right but what is going wrong and needs to be improved. He also recognizes that assessment or appraisal may be less than objective, political in its choices, and ineffective in its outcomes, for a range of reasons such as those outlined below.

Performance appraisal by line managers or other supervisors as a means of raising productivity may be limited by the following factors [73]:

- belief that accurate ratings would have a damaging effect on subordinate motivation and performance;
- desire to improve the subordinate's chances of getting a pay rise;
- a wish to prevent a superior from obtaining evidence of internal problems and conflicts;
- prevention of a permanent written record of poor performance that might have longer-term implications for the subordinate;
- need to protect subordinates whose performance suffered from the effects of personal problems;
- desire to reward subordinates who had put in a lot of effort, even if the end result was not so good;
- avoidance of confrontation and potential conflict with "difficult" subordinates;
- desire to promote subordinates from outside the department who were disliked or problem performers;
- desire to scare people into performing better;
- punishment for difficult or noncompliant subordinates;
- encouragement of unwanted subordinates to leave; and
- minimization of merit pay rewards.

One of the responses in principle to this is a 360 ° evaluation with subordinates directly evaluating superiors or peers as well as being subject to review by them. Yet this also may be risk-prone, and not least for the individuals concerned. Who can be sure that the confidentiality of the assessment will be secure? Who wants to call a colleague, especially a superior, incompetent if the result of the exercise will be publicized? And if it won't be, what is the point?

A framework for performance assessment that enabled employees to talk about what is going right, not right, or seriously wrong and make recommendations for remedying the situation could feed up to higher levels of management and indicate what is dysfunctional rather than simply whether the employees have performed according to criteria that could actually be the root cause of dysfunction in the organization. Such dysfunctional criteria might result in workers feeling alienated and might not lead to greater social efficiency for the public.

Nor is a 360 ° evaluation in the sense of assessment by both an employer (or a line manager or supervisor) and an employee inclusive when it is done on a bilateral and thus dyadic basis alone. It excludes "externalities" related to what employers should be able to recognize in terms of not only work but also life experience in employees, such as family or other personal reasons why someone may not be performing well at work. In many cases, this will have more to do with the personal life of workers rather than only work performance. X may have been a good worker, or manager, sensitive to colleagues and the needs of the job, or work group, or whatever. But then some life crisis happens, such as the breakdown of a marriage or other personal relationship, or a close family illness, or perhaps the parent of an employee was diagnosed with terminal cancer and only has a few months to live, which may mean that he or she needs "time off" to be with and care for them. But the worker may not have the right to take a leave since it has not been stipulated in his or her employment contract.

4.3.4 Economic versus social efficiency

One of the main limitations of NPM is that its organizational logic has been designed from the top down without concern for learning up from organizational levels on what works well, what does not, and what could be made right by more efficient use of time and resources. For instance, a European teaching hospital in one of our case studies [62] consistently ranks high in the OECD for the quality of its medical care. Medical and nursing staff are highly flexible and multitask in a manner also consistent with the operational logic of a Toyota style post-Fordism. But the organizational logic of the hospital is "several things at the same time" with multiple inefficiencies, most of which remain unresolved because they have not "surfaced" through dialog and voice for health professionals at operational levels:

- First, its formal organizational structure is multidivisional as in a classic Sloan-GM model, with each department or service having its own vertical structure.
- Second, within the departments or services authority is hierarchical, but in a pyramidic Weberian sense, with all real authority focused at the apex of the departments, while the wider base of the 4,000 medical staff and nurses have no authority to take initiative or feed up operational learning.
- Third, there is no mechanism for lateral exchange of learning by experience through relational coordination [74] and boundary spanning [75] either between departments or, often, within them.
- Fourth, surgical teams within units under pressure to increase patient throughput tend not to have time to share their operational learning with other teams, while even this tends to be casual and learning only on an implicit double-negative basis by chatting and anecdote about incidents in which things had gone wrong.
- Fifth, it is difficult for the management board of the hospital, including its trustees, to know much of what is happening at the consultant level or below within individual departments and services, or obtain a clear idea of what could be gained from this by learning up from operational logic for organizational logic [25, 62, 63].

This was not because the hospital management was not concerned with inertial institutional logic [20]. Nor was this only externally by pressure on the government from the Troika of the IMF, European Central Bank, and European Commission to reduce costs. Although this was a factor, it was secondary to its primary concern that the hospital was not functioning at anything like its potential in terms of service, while the high degree of alienation of already overworked nursing staff at the base of its organizational pyramid was resulting in levels of absenteeism that were seriously compromising operational efficiency. This was aggravated by the concern of the government to introduce Taylorist criteria for performance, which were increasing the intensity of work and stress.

More dramatically, 2016 saw the first all-out strike action by junior doctors in the English National Health Service (NHS) since the NHS was founded in the UK in 1948. This was in response to a government demand for a new contract obliging them to work at cheaper rates on weekends – something ministers claimed was needed both to improve care on Saturdays and Sundays and to reduce costs. But it also represented an intensification of the labor process, and thereby stress, and a disregard for work-life balance. The British Medical Association supported the junior doctors on the grounds that the government should both pay either as much or more for week-end work and hire more doctors to do it.

When polled, 99% of junior doctors voted in favor of industrial action short of a strike, and 98% for full strike action, demonstrating the strength of feeling in the profession. Industrial action went ahead in January, with juniors withdrawing their labor and providing only emergency care. The government did not respond. The doctors then took part in their second all-out strike in April 2016 in protest over the imposition of a new contract. The April strikes were the first-time doctors had stopped providing emergency care in the history of the NHS. By August 2016, when the new contract should have come into force, the dispute had not been resolved [76, 77].

Thus what transpired was a contradiction of social efficiency in the sense of the wellbeing of both medical staff and patients by an unremitting demand for greater economic efficiency. This concerns the need to appreciate what have been identified as proximal and distal processes and outcomes [60, 78]. The proximal outcomes (e.g., absenteeism, turnover rates, productivity, quality, and service performance) relate more closely and directly to human resource (HR) practices (e.g., selection, training, motivation, and opportunity to contribute). Distal outcomes relate to the overall financial or market performance of an organization (e.g., return on capital and market value if it is in the private sector, or how much care costs in a NHS) and may appear to senior managers to be less directly linked with HR practices. Yet the one can either aid or abet the other, which justifies more attention to not only gains but also losses within a HPWS rather than a more relational approach [15] within a HIWS.

4.4 Hierarchy, surveillance, and education

The introduction of primary education in the later nineteenth century was based on what businesses needed were people who could read, write, calculate, and perform rather than imagine, create, or challenge. For Cherkaoui [79], the forms of assessment that emerged with mass education were a mode of socialization that prepared individuals for the division of labor within a Weberian bureaucratic society. For Foucault 07], this was part of the "normalization" required from modern education itself and, with it, surveillance to ensure that norms were observed. This was echoed by Bourdieu [80] in his *Homo Academicus*, which also asserted that norms needed to be respected rather than challenged if one wished for advancement.

Most mass education not only is Fordist in its concern for volume throughput, Weberian in terms of hierarchy and Taylorist in terms of explicit rules for achieving standards. It is also Foucauldian in knowledge as power and power in controlling the reproduction of knowledge, including what research is deemed worthy of government support. This was well put in a debate on a proposed reform of Oxford University by one of its fellows teaching management, Peter Johnson [81]. The proposal was to appoint a management board over and above the "congregation" of the college fellows who actually did the teaching, with a chief executive officer and an inner group who could make key decisions on supposedly clear-cut criteria. As Johnson put it:

> the particular business variant underlying the new governance proposals is well past its sell-by date. The engineers' paradigm behind [them] applies the ideas of Henry Ford and Frederick Taylor to learning as though the university were a machine bureaucracy where dons are substitutable employees in an integrated academic factory. [81, p. 15]

Misplaced market criteria in education in the UK also have included the introduction of private sector academies to replace local government responsibility for secondary schools. Launched initially by Margaret Thatcher as City Technology Colleges, they presumed that public-spirited business leaders would sponsor local schools, impose higher standards, and, by linking business more directly with education, foster entrepreneurship in young people. Taking "market solutions" seriously, one city academy started to pay pupils cash bonuses to achieve target exam results, even if the expected grade was the lowest to get a pass rather than a fail [82].

Overall, academies have removed Hirschman's [83] "voice" for local governments and parents, yet not improved examination performance. A recent report from the Education Select Committee of the House of Commons on Academies and Free Schools [84] concluded that academy status does not result in raised standards, that schools work best in collaboration with others not in isolation, that sponsors and proposers of academies and free schools have not been properly vetted before being allowed to run taxpayer-funded schools, and that the whole system lacks transparency as well as oversight and is open to fraud, abuse, and mismanagement.

Clearly there are dilemmas here. At one level, tests and assessment are important as indicators of whether someone is competent in terms of explicit reasoning. Yet too much testing, especially within schools, can inhibit creativity, especially when it is done according to "national standards" and when schools are assessed for future funding on a productivity basis of how many examinations per pupil have been passed at what grades. This risks meaning less the *educare* as leading out that Rousseau stressed in his *Emile or on Education* [85] than the *inducare* or induction into what Alfred North Whitehead [86], coauthor with Bertrand Russell of *Principia Mathematica*, deemed too often to be "dead bodies of knowledge" and into the technique of how to pass examinations rather than Whitehead's concern that education should be about "life itself."

The potential positive effects of HR practices on organizational performance in education depend on student/staff-related factors (e.g., individual or group wellbeing in relation to those teaching courses or supervising research). This implies transcending an AMO framework [52, 87, 88]. It also differs from human capital models of HRM such as HPWSs [14]. By contrast, relational models of HRM go further than the AMO framework by strengthening relationships between students and staff [15] and enhancing a high-involvement learning environment.

Yet teachers as the line managers of mass education know that if they fail to show productivity in terms of both the quality and volume of examination passes, both those they are teaching and they themselves will be sanctioned. They know that they cannot afford to encourage imagination and diversity in the answers that their pupils give on tests since, if they do, both the pupils' ratings and theirs as assessors will drop because the pupil – or student – has departed from a top-down assessment "template," and their "internal market" rankings will fall.

In addition, despite the admirable aim of European education ministers to achieve diversity and quality in European higher education in their Bologna Declaration [89], the Bologna Process adopted to achieve this neglected quality and introduced a crude quantitative input-output model in its assessment procedures. It ranks universities in terms of the input of hours taught, yet pays no attention to class size. It insists that a university must allocate a given number of hours to teaching to earn a high ranking by its own criteria. This neglects entirely the notion that the value of teaching may be *inversely* related to class size and reflects both a constrained and flawed concept of productivity [90].

For example, a graduate class of 12 is one in which the teacher can get to know not only the names of the students and what they can do well or are doing less well, but also who they are, what their backgrounds are, what their aspirations are, in particular, those related to careers and quality of working life. In a class of 24, this is still possible. In a class of 48 it becomes geometrically rather than only arithmetically more difficult. In an undergraduate lecture of 240 students, it is impossible. Assessment of the quality of teaching in terms of input hours therefore is absurd if one wishes to evaluate the quality of teaching irrespective of class size. The Bologna Process pays no attention to this consideration nor does it rank it in its assessment procedures.

4.5 Counterproductive health reforms

From the 1980s the Thatcher and Blair governments in the UK did not overtly aim to privatize the NHS, and they did not commit themselves to doing so in their election manifestoes. But their presumption was that a flexible private sector is always more efficient than inflexible public sector "bureaucracies." Their explicit logic was the introduction of an "internal market" to the health service and outsourcing of services that hitherto had been internal to hospitals or health care centers. It was claimed that this would raise quality and widen freedom of choice [91, 92]. Its managerial logic was that with more outsourcing and shorter-term internal contracts, the power of professional associations and trades unions in the NHS would decrease [13]. Its implicit economic logic was Fordist in seeking to gain higher patient throughput and Taylorist in terms of introducing line managers to assure that new performance criteria were being observed by health care professionals.

Yet this proved entirely counterproductive in terms of economic efficiency. Many of the new line managers had no medical training. The top-down performance criteria were decided by civil servants who, with very few exceptions, likewise had no medical training. The outcome was entirely counterproductive. It trebled administrative costs in the British NHS, and the English NHS after devolution of health services to Scotland, Wales, and Northern Ireland, from less than 5% in the 1970s to 14% by 2004 [23], which, despite government claims that this would be reduced by 2011, was not lowered because of the inertial top-down organizational design of the Nep Public Management paradigm [22].

British shadow finance minister Gordon Brown had made a commitment before the 1997 general election that a Labour government would not breach the budget limits of the previous Conservative government, including those for the NHS, which appeared Thatcherite. Yet then, with Labour winning the next general election, in a very un-Thatcherite manner, he threw money at the NHS, nearly doubling, for example, the income of local doctors or general practitioners and increasing funding for local health authorities and hospitals [22].

The outcome of more money, and more health care employees, had been a reduction in waiting lists. But the ongoing pursuit by government of higher Fordist volume on the presumption of economies of scale, rather than of scope, also posed "size" problems. Some of the chief executives of new local and regional hospital "trusts" chose to take early retirement rather than manage trusts that the government enlarged, and then enlarged again, since in their view they then would have been too big for anyone to know what was going on, much less continue with "change management by consent" [36].

In addition, although Gordon Brown had prided himself on the principle that his "prudent" public management of finances would end an earlier cycle of "boom and bust," this was the outcome for new funding without a new management model for the NHS since, while increasing resources for it overall, he also imposed strict penalties if they were exceeded. For, in trying to meet local demand, some hospitals ran out of money, as a result of which not only did many have to cut some services, but others closed outright. This led to the resignation of the chief executive of the NHS but without any recognition of what had gone wrong in terms of management rather than choosing simply to quit in view of the resulting crisis [93]. Moreover, Bolton [94] has observed that at the heart of attempts to gain higher productivity in health within a NPM paradigm, there is an illogic that emphasizes contradictory criteria in claiming both to improve the quality of service yet also cut those who could deliver it and that "As nurses account for the largest part of the hospital budget, and also are accountable for how the quality of bedside care is perceived, these contradictions deeply affect their work" (94, p. 320).

The new Taylorist focus on internal hospital efficiency also neglected resulting social diseconomies with a degree of deceptive accounting verging on little less than deceit. For example, part of the increase in patient admissions was due to their spending less time in the hospital and, thus, the outsourcing of postoperative care from hospital staff to general practitioners or health workers, or families, whose members might have to take time off work to care for those who had been sent home without having fully recuperated. Faster throughput also meant a higher rate of readmissions of patients who formerly would have stayed longer in hospital. Perversely also, although suiting government claims on admissions, the total figures for these included those who had to be readmitted because of complications or failure to recuperate, with Sterns' "revolving door" effect that sending patients home early raised "success" in later readmission rates [95].

4.6 Logic in learning from lean

By contrast with these failures in change management within an unreconstructed Weberian and Fordist paradigm in the UK, other governments were learning up from lean management on the post-Fordist flexible production model [30, 96]. Teich and Faddoul [97] found that most of these were in the USA but that early approaches implementing lean principles were not much more than an exercise in reducing stock inventories in hospitals. Nonetheless, later approaches included managerial and support case studies, patient-flow case studies, and organizational case studies.

Teich and Faddoul give several examples of the successful implementation of comprehensive lean projects in health care in the USA, such as at the Virginia Mason Medical Center, where the hospital reported increased profit margins, decreased deaths, and a reduction in the number of medication errors. Other reported benefits were an 85% reduction in how long patients had to wait for lab results and a lowering of inventory costs by $1 million. In order to directly learn up from lean production systems, in 2002 30 senior managers traveled to Japan for 2 weeks to observe this at Toyota. By 2008, more than 200 employees visited production plants in Japan [97].

Researchers from Belgium explored whether lean management activities improved patient safety culture in a radiotherapy institute. Data were collected over a 3-year period using surveys, workshops, an incident reporting system, and interviews with professionals. Lean approaches were associated with some changes in safety culture and increased intention to take action to prevent future incidents. The number of patient safety incidents decreased owing to better group-based problem solving and fuller use of resources. Patient safety culture improved significantly as a result of the introduction of so-called care pathways [97], which is consistent with Gittell et al. [74] on relational coordination and Mørk et al. [75] on boundary spanning.

Flexible capacity utilization, care pathways, and concern with cutting waiting times were among the most notable examples of a post-Fordist reorganization on lean principles in the Karolinska teaching hospital in Stockholm in the 1990s, inspired by its general director Jan Lindsten [32, 98]. Flexible capacity use of operating theaters was introduced based on the likely duration of the operation rather than the type of operation.

As a result, all theaters were segmented into four groups – fast, medium, slow, and emergency. Flexible theater use was matched by a flexible use of wards. Underlying this was the principle of clustering so-called families of service both within the organizational logic of the hospital and the operational logic of departments and units. In addition, a new post of nursing Coordinator was created and adopted by most departments. This job's responsibilities include minimizing the number of visits a patient must make and scheduling preoperative preparation and postoperative care by doctors. Two operating theaters were closed, but a new preoperation anesthetic clinic opened. More anesthetists were hired, allowing senior staff to focus on surgery and junior staff to manage all preoperative patients.

The outcomes were that the new position of Nursing Coordinator created a career path for nurses, who in practice became responsible for the administration of the various departments. This freed doctors from administration and allowed them to spend more time on their clinical work and research. The time between operations was cut by up to half. Average waiting time for surgery was reduced for hip operations from 8 months to 3 weeks and for some cancer conditions, after diagnosis, to 3 days. Overall unit costs were cut by 15%. No medical staff were made redundant.

4.7 Responding to technological unemployment

The labor-substituting effects of technical progress, and thus technological unemployment, are now attracting increased attention. Some of this ranges from an optimistic scenario of increased welfare [99] to cataclysmic in its pessimism. Thus Frey and Osborne, of the Oxford University Martin Programme on the Impacts of Future Technology, anticipate that up to 47% of jobs in the USA could be displaced by computer technology and robotics within the next two decades. Analyzing some 702 different occupations, they found that the jobs most at risk were semiskilled or unskilled, which has major social implications [100].

Mainstream macroeconomics tends to assume that technical progress is neutral. But it is not. It has differential economic and social implications. As with horses with the advent of the horseless carriage, in the case of the driverless car being developed by Google, jobs for hundreds of thousands to millions of commercial drivers of cabs and freight vehicles could disappear. Larry Page of Google has claimed that new technologies will make businesses not 10% but 10 times more efficient and that, provided this flows through into lower prices, "I think the things you want to live a comfortable life could get much, much, much cheaper" [101, p. 4].

Yet if there is to be an increase in welfare and wellbeing, this demands that the productivity increases be distributed within society in terms of social income, social investments, or social expenditures by governments in ways that compensate for the job losses involved. For example, while technical progress is still creating jobs in firms, it is creating fewer of them. In 1960, the most profitable company in the world's largest economy was GM. In today's money, GM made $7.6 billion that year. It also employed 600,000 people. Today's most profitable company, Apple, employs 92,600. Where 600,000 workers once generated $7.6 billion in profit for GM, 92,600 in 2014 generated $89.9 billion for Apple, an improvement in profitability per worker of 76.65 times [99].

Technological unemployment is consistent with Piketty's findings [69] on inequality but also gives more explanatory power in terms of asymmetric outcomes from technical progress than the neutrality assumed in mainstream macroeconomic theory. It also implies more than Piketty's proposals [69] for global taxes on wealth and income since, through transfer pricing, which is the near exclusive domain of multinational corporations rather than smaller national firms, vast profits by companies such as Apple, or Google or Starbucks are not being effectively taxed.

On this point, Piketty makes no proposals on how to achieve more effective taxation. There had been such proposals made at high levels in the Commission to Jacques Delors when he was its president, and they attracted his support and that of then director general of Eurostat [2, 102]. But this support lapsed after Delors resigned in 1995 following his 10-year tenure as the commission president. After which, multinational corporations were highly effective at bringing pressure to bear on the subsequent president of the commission, Jean-Claude Juncker, to arrange for them tax avoidance deals when he was prime minister and finance minister of Luxembourg [103]. As Richard Murphy, a tax expert and campaigner who runs Tax Research UK, told International Business Times, UK:

> The whole of Europe faces a financial crisis [. . .]. It's going into recession. Its member states have one characteristic in common; they are short of tax revenue. And yet, as prime minister of Luxembourg, Juncker made it his job to ensure that, frankly, a blind eye was turned to tax avoidance across Europe being facilitated by his Duchy. [103]

Yet the combination of low incomes, current unemployment, and social and political tensions arising from both already have been sufficient to change the terms of reference in politics in the USA, with Donald Trump and his assault on globalization decimating other candidates for the Republic nomination for the presidency of the USA as well as in the support for the self-styled democratic socialist Senator Bernie Sanders in the primaries leading to the adoption of Hillary Clinton as the Democratic candidate.

This had echoes also in the UK in terms of the 2016 Brexit vote. In Paul Mason's view [11], shared by one of us with extensive experience of British politics, this was less a resurgence of nationalism – any more than the remarkable success of the Scottish Nationalist Party in the previous general election – than a protest against a neoliberal model of politics that prioritized market rather than social values. It also had precedents in earlier referendums in Europe for the three countries given the chance to vote on the ill-fated constitution for Europe proposed by former French President Valéry Giscard d'Estaing, France, the Netherlands, and Ireland, all of which voted against. On the other hand, opinion polls showed that the constitution was rejected not because most people at the time were "against Europe." For instance, a Gallup poll in France in June 2005 found that 83% of those voting against thought that "EU membership is a good thing"; the same share also thought that voting "No" would give the "opportunity for a more social Europe," while 80% wanted a treaty that would "better defend national interests and jobs" [104].

The answer to both technological unemployment and the outsourcing of jobs with globalization is not competing down in terms of incomes and rights at work but more labor-intensive employment in the social domains of health, education, and public and social services. In addition, such employment overwhelmingly is local not global, not exposed to international competition and not necessarily subject therefore to the efficiency criteria of the private sector. Finally, as also touched on at the outset, to the degree to which job creation in this field is public rather than private, its employees are in the "overground" economy and taxed at the source, with multipliers in terms of income and employment that then generate demand in the private sector.

4.8 Working to mutual advantage

What has been represented so far is that, in the social domain of health care provision in the UK, "change management" focusing in a Fordist manner on throughput, with Taylorist surveillance, has not delivered expected gains from NPM reforms, while continual surveillance and assessment of students on the grounds of needing to compete in a global era is widely held by the UK teaching profession to have degraded the quality of education. The last people to have been consulted in both cases were those on the front lines in health care provision and teaching, whereas it could have been to the mutual advantage of both parties and policymakers to have involved them from the outset in what change was feasible to increase both economic and social efficiency and in what ways.

However, unless there are procedures in place for recognizing proposals for innovative methods of work operation and share their added value with an organization or institution, few employees, including middle management, will be motivated to make such proposals to increase either economic or social efficiency. This has several implications:

- Employees proposing innovative methods of work operation, including middle managers, must know that they are not thereby innovating themselves or colleagues out of a job.
- Not every organization can commit to profit sharing on the model of the Toyota Production System, not least nonprofit organizations, such as most public services. Not everyone can be promoted or paid more than efficiency allows. But striking a work-life balance through being able to customize individual working time is a key motivator.
- Multiskilling and multitasking may not reduce stress, but skill profiling and skill path planning [36] can enhance individual fulfilment, while horizontal mobility can both provide job variation and lessen the incidence of intensive "front line" work pressure.
- Innovations in methods of work operation focused on cutting wasted time and better use of fixed resources are a key alternative to raising productivity rather than to longer hours or cutting jobs. Not every organization can innovate new products, but all can in principle innovate new methods of work operation and thereby improve operational efficiency.
- The best resource for achieving this within an innovation ideology rather than only a cost-cutting ideology is employees themselves. It is they who know best the tacit rules and implicit norms that are frustrating operational and organizational learning.
- Effective mutual feedback is vital if both management and employees are to be able to voice [83] not only their own interests but also their earlier learning from experience on how operational logic can be improved. Such voices should be able to articulate and improve what otherwise is implicitly assumed or has been eroded.

4.8.1 Innovation-by-agreement and flexibility-by-consent

Such a mutual advantage approach or paradigm was advocated by one of us to the then prime minister of Portugal Antonio Guterres in recommending innovation-by-agreement or flexibility-by-consent [38] and has four main implications for countering social inefficiency from excessive pressure for economic efficiency:

- First, pressure to increase throughput in a Fordist manner may increase productivity, but it may also increase fault rates and deprive people of the time needed either for reflective practice [46–48] or learning up from operational levels on the potential for continuous improvement through reducing wasted time and materials [29].
- Second, if increasing throughput means increased overtime, this will both will raise unit costs and tend to be unsustainable in terms of work-life balance, unless there is agreement with employees to credit current overtime against later "undertime."
- Third, while few Western companies exposed to global competition are going to commit themselves to lifetime employment on the Japanese model, they can commit themselves to no-redundancy agreements for the expected life cycle of a product or model; this could have the same incentive for employees to commit themselves to continuous improvement and operational learning.
- Fourth, a socially embedded mutual advantage paradigm at the operational level may achieve flexibility-by-consent and sustainable continuous improvement, and thus operational learning, through reinforcing rather than reducing employee rights, as was the intent and design of the innovation-by-agreement approach of the Lisbon Agenda.

The right to negotiate working time to enhance personal work-life balance was explicitly recommended in the Lisbon Agenda agreed in 2000 by the European Council of heads of state and government [38, 105]. It could be combined in the public sectors of education and health with progress on more labor-intensive employment in the social sphere [102], which was endorsed by the Essen European Council [106], with more teachers and smaller classes, more health care workers, and shorter waiting lists. Continuous improvement negotiated through innovation-by-agreement therefore need not be limited to the production sphere or private services. It can include the following:

1. the right of workers and managers in both the private and public sectors to expect negotiation to range beyond wages and working conditions and to include the relation between their work and nonwork lives as well as retraining, job redesign, skills path planning, and horizontal mobility;
2. the degree to which this and the personalization of service and "continuous improvement" in education, health, public administration, and other public services can directly benefit the wider public, enhance social efficiency, and improve the quality of life.

Innovation-by-agreement as a process, based on flexible production rather than flexible labor markets, offers efficiency both within organizations and for society. It can achieve a positive-sum institutional logic in terms of new methods of work operation based on consent because the process reinforces individual rights. But it also can imply a positive external logic for society as a whole. By allowing employees more scope from saving time through new methods of work operation, innovation-by-agreement offers both the potential for better work-life balance and for employees to be able to relate better to individual patients, students or those claiming benefits.

4.9 Summary

- An efficient market is concerned with economic criteria, competitive advantage, and private gain. An efficient society is concerned with social criteria, mutual advantage, and social gain.
- An efficient economy is concerned with market innovation. An efficient society is concerned with social innovation, as was the case with the innovations in national insurance and the right to a pension, or NHSs or of public and mutual societies for housing.
- An efficient market will meet consumer preferences. An efficient society will meet social preferences, such as for better health, education, and quality of life and the environment.
- An efficient economy counters unemployment by effective demand. An efficient society will match latent social demand with effective supply.
- An efficient society will be concerned with both full and socially useful employment.
- An efficient society will recognize that not everyone in an economy has to be efficient or hyperefficient by the highest standards in the market domain. Part of an economy may be so and sustain the rest well at high levels of employment.

Japan is an exemplar of the last point, in which less than a seventh of its employment has been in hyperefficient industrial groups, with commitment to continuous improvement in methods of work organization to raise productivity. The rest of the economy has low productivity in both agriculture and services yet is socially efficient in the sense of assuring employment, income, and a high degree of social cohesion [28, 107]. The low growth that has followed in the wake of Japan's own financial crisis in the 1990s is less negative than positive in being less damaging to the environment in a society that already has achieved some of the highest living standards in the world.

References

[1] Lacey R (1987). Ford: The Men and the Machine London: Heinemann and Pan Books.

[2] Holland S (2015). Europe in Question – And What to Do about It Nottingham: Spokesman Books.

[3] Friedman M (1980). Free to Choose New York: Harcourt Brace Jovanovitch.

[4] Hodges A (1983). Alan Turing: The Enigma London: Vintage.

[5] Mazzucato M (2011). The Entrepreneurial State London: Demos.

[6] Schumpeter J (1911). 1949 The Theory of Economic Development Cambridge MA: Harvard University Press.

[7] Foucault M (1977). Discipline and Punish London: Allen Lane.

[8] Denison EF (1967). Why Growth Rates Differ Washington: The Brookings Institution.

[9] Carroll CD, Slacalek J, Tokuoka K (2014, March). The distribution of wealth and the marginal propensity to consume. ECB Working Paper 1655. https://www.ecb.europa.eu/pub/pdf/scpwps/ecbwp1655, last accessed in 2014.

[10] Holland S (2016). Beyond Austerity – Alternatives for a Democratic Europe Nottingham: Spokesman, 15 March 156.

[11] Mason P (2015). Postcapitalism. London: Allen Lane.

[12] Keynes JM (1936). The General Theory of Employment, Interest and Money London: Macmillan.

[13] Le Grand J (1997). Knights, knaves or pawns? Human behaviour and social policy. The Journal of Social Policy 26(2): 149–169.

[14] Crook TR, Todd SY, Combs JG, Woehr DJ, Ketchen DJ (2011). Does human capital matter? A meta-analysis of the relationship between human capital and firm performance. Journal of Applied Psychology 96(3): 443–456.

[15] Gittell JH, Seidner R, Wimbusch J (2010). A relational model of how high-performance work systems work. Organization Science 21(2): 490–506.

[16] Varoufakis Y, Holland S, Galbraith JK (2014). Modeste Proposition Pour Résoudre la Crise de la Zone Euro. Preface Michel Rocard Paris: The Veblen Institute and Les Petits Matins, January.

[17] Lindbeck A, Snower DJ (1988). The Insider-Outsider Theory of Employment and Unemployment Cambridge, MA: MIT.

[18] Day D, Miscenko D (2016). Construct Evolution, Contributions, and Future Prospects for Advancing Leadership Theory. In: Leader-Member Exchange Theory. Bauer T, Erdogan Berrin B (eds) NY: Oxford University Press, 9–28.

[19] Ilies R, Nahrgang JD, Morgeson FP (2007). Leader–member exchange and citizenship behaviors: A meta-analysis. Journal of Applied Psychology 92(1): 269–277.

[20] Lok J (2010). Institutional logics as identity projects. Academy of Management Journal 53(6): 1305–1335.

[21] Oliveira TC, Holland S, Fontes da Costa J, Lira de Carvalho FE (2014). Hierarchies or Holdings?. In: Effective Human Resource Management in Small and Medium Enterprises: Global Perspectives, Machado C, Davim JP (eds), IGI Global Associates, 342–376.

[22] Leys C, Player S (2011). The Plot against the NHS London: Merlin.

[23] Pollock AM (2004). NHS Plc London: Verso Books.

[24] Janssen R (2015). Labour Market Deregulation and Productivity – IMF Finds No Link, Social Europe, April.

[25] Oliveira TC, Holland S, Filipe NC (2016). How to Learn up from Lean Management in Health Services? HRM, Leadership and Relational Coordination. In: Lean and Green Management, Machado C, Davim JP (eds), Springer, 109–137.

[26] OECD (1999). Employment Protection and Labour Market Performance Paris: Organisation for Economic Cooperation and Development.

[27] Esping-Andersen G, Reginin M (eds) (2000). Why Deregulate Labour Markets? Oxford: Oxford University Press.

[28] Moriguchi C, Ono H (2004, May) Japanese lifetime employment: A century's perspective European institute of Japanese studies, EIJS Working Paper Series (No. 205). Retrieved from https://ideas.repec.org/p/hhs/eijswp/0205.

[29] Colenso M (2000). Kaizen Strategies for Organisational Change: Evolution and Revolution in the Organisation London: Financial Times-Prentice Hall.

[30] Womack JP, Jones DT, Roos D (1990). The Machine That Changed The World New York: Harper Collins.

[31] Ohmae K (1982). The Mind of the Strategist: The Art of Japanese Business New York: McGraw-Hill.

[32] Kaplinsky R (1995). Patients as Work in Progress: Organisational Reform in the Health SectorIn: Europe's Next Step: Organisational Innovation, Competition and Employment, Andreasen L, Coriat B, Den Hertog F, Kaplinsky R (eds) London: Frank Cass, 286–304.

[33] Guest DE (1998). On meaning, metaphor and the psychological contract: A response to Rousseau. Journal of Organisational Behaviour 19(1): 673–677.

[34] Rousseau DM (1998). The problem of the psychological contract. Journal of Organisational Behavior 19: 665–671.

[35] Rodrigues R, Guest D, Oliveira T, Alfes K (2015). Who benefits from independent careers? Employees, organizations or both?. Journal of Vocational Behavior 91: 23–34.

[36] Holland S, Oliveira TC (2016). Flexible labour, flexible production and innovation-by-agreement: International comparisons contesting the Lindbeck-Snower insider-outsider thesis and 'structural reforms' in the European union European Journal of Comparative Economics 14(1):89–107.

[37] Yamaguchi K (2004). The state and change in the 'lifetime employment' in Japan: From the end of war through 1995. Keizai Sangyo Journal 11.

[38] Holland S (2000). Innovation Agreements. Paper for the Portuguese Presidency of the European Council Lisbon: Office of the Prime Minister.

[39] Weber M (1947). The Theory of Social and Economic Organisation (Part 1 of Wirtschaft Und Germeinschaft) Parsons T (ed) London: William Hodge.

[40] Lunenburg F (2010). Leader-member exchange theory: Another perspective on the leadership process. International Journal of Management, Business and Administration 13(1): 1–5.

[41] Chen Z, Lam W, Zhong JA (2007). Leader-member exchange and member performance: A new look at individual-level negative feedback-seeking behavior and team-level empowerment climate. Journal of Applied Psychology 92(1): 202–212.

[42] George JM, Jones GR (2008). Understanding and Managing Organizational Behaviour (5th edn). Upper Saddle River, NJ: Prentice Hall.

[43] Bass BM (1985). Leadership and Performance beyond Expectations New York: The Free Press.

[44] Bass BM (1999). Two decades of research and development in transformational leadership. European Journal of Work and Organizational Psychology 8(1): 9–32.

[45] Oliveira TC (2007, June). Delving down to Learn Up. Plenary Presentation to the Sloan MIT-Portugal Conference on New Developments in Management, Lisbon, Portugal.

[46] Schön DA (1983). The Reflective Practitioner New York: Basic Books.

[47] Schön DA (1987). Educating the Reflective Practitioner San Francisco CA: Jossey-Bass.

[48] Schön DA (1991). The Reflective Turn New York: Teachers College Press.

[49] Holland S (2013). The Trajectory of the Periphery. Contribution to the Conference on Can the Eurozone Be Saved? Lyndon B Johnson School of Public Affairs, Austin Texas, 4 November.

[50] Matthijs M, McNamara K (2015). The Euro crisis. theory effect: Northern saints, Southern sinners, and the demise of the Eurobond. Journal of Economic Integration 37(2): 229–245.

[51] Oliveira TC, Holland S, Marques Leitão M (2003). Site Visits to and Direct Enquiry from Volkswagen Autoeuropa Palmela, Portugal: Volkswagen Autoeuropa Palmela.

[52] Appelbaum E, Bailey T, Berg P, Kalleberg A (2000). Manufacturing Advantage: Why High–Performance Work Systems Pay Off Ithaca, NY: Cornell University Press.
[53] Oppel E-M, Young G (2016). Nurse Staffing Patterns and Patients' Perceptions of Quality: An Empirical Analysis of U. S. Hospitals. Paper Presented at the European Health Economics Association (Euhea) Conference. Hamburg, July.
[54] Everhart D, Schumacher JR, Duncan RP, Hall AG, Neff DF, Shorr RI (2014). Determinants of hospital fall rate trajectory groups: A longitudinal assessment of nurse staffing and organizational characteristics. Health Care Management Review 39(4): 352–360.
[55] Spetz J, Donaldson N, Aydin C, Brown DS (2008). How many nurses per patient? Measurements of nurse staffing in health services research. Health Services Research 43(1): 1674–1692.
[56] Mark BA, Harless DW, McCue M, Xu Y (2004). A longitudinal examination of hospital registered nurse staffing and quality of care. Health Services Research 39(2): 279–300.
[57] Brosig-Koch J, Kairies-Schwarz N, Kokot J (2016). Sorting into Physician Payment Schemes – A Laboratory Experiment. In: Paper Presented at the European Health Economics Association (Euhea) Conference, Hamburg, July.
[58] Douven R, Remmerswaal M, Zoutenbier R (2016). Do altruistic mental health care providers have better treatment outcomes? CPB Bureau for Economic Policy Analysis July 1–34.
[59] Schneider U, Schmid A, Verheven F (2016). Treatment Decisions and Hospital Organization: Do Owner and Type of Ownership Matter? an Analysis of the Choice of TAVI for Patients with Aortic Stenosis. In: Paper Presented at the European Health Economics Association (Euhea) Conference, Hamburg, July.
[60] Peccei R, Voorde K, Veldhoven M (2014). HRM; Well-Being and Performance: A Theoretical and Empirical Review. In: HRM and Performance: Achievements and Challenges, Paauwe J, Guest D, Wright P United Kingdom: John Wiley & Sons Ltd, 15–46.
[61] Oliveira TC, Holland S (2007a). Alternative paradigms of hospital work organisation and health provision. Revista Portuguesa de Saúde Pública 25(1): 19–37.
[62] Oliveira TC, Holland S (2013). Institutional Logics Promoting and Inhibiting Innovation. In: Management and Engineering Innovation, Machado C, Davim JP (eds), IGI Global Associates, 179–216.
[63] Oliveira TC, Raposo V, Holland S, Lira de Carvalho FE (2016). Hospital Governance and Health Reforms: Challenges for Lean and Hybrid Managers. In: Lean and Green Management, Machado C, Davim JP eds, Springer, 161–187.
[64] Kregel J, Matzner E, Grabher G (1992). The Market Shock Ann Arbor: Michigan University Press.
[65] Boyes R (2006). New Europe, old dangers The New Statesman October, accessed on 15th July 2025 from https://www.newstatesman.com/long-reads/2006/10/new-europe-old-dangers.
[66] Wagstyl S (2006). Sick men of new Europe: Why reform fatigue has hit the East The Financial Times 27 September.
[67] Seddon J, O'Donovan B (2010). Efficiencies in Public Service Delivery. In: From Recession to Renewal: The Impact of Financial Crises on Public Services. Richardson J (ed) Bristol: Policy Press, 135–156.
[68] Blond P (2009). The Ownership State London: Res Publica/NESTA.
[69] Piketty T (2014). Capital in the Twenty-First Century, Belknap and Harvard University Press.
[70] Barber M (2007). Instruction to Deliver London: Portico's Publishing.
[71] Barber M (2015). How to Run a Government so that Citizens Benefit and Taxpayers Don't Go Crazy London: Allen Lane.
[72] The Economist (2015). Delivery Man. Tony Blair's "Deliverology" Expert Explains How to Reform Public Services 28 March, accessed on 15th July 2025 from https://www.economist.com/books-and-arts/2015/03/26/delivery-man.
[73] Fletcher C (2001). Performance appraisal and management: The developing research agenda. Journal of Occupational and Organizational Psychology 74(4): 473–487.

[74] Gittell JH, Weinberg D, Pfefferle S, Bishop C (2008). Impact of relational coordination on job satisfaction and quality outcomes: A study of nursing homes. Human Resource Management Journal 18(2): 154–170.
[75] Mørk BE, Hoholm T, Maaninen-Olsson E, Aanestad M (2011). Changing practice through boundary organizing: A case from medical R&D. Human Relations 65(2): 263–288.
[76] BBC News (2016). Junior Doctors' Strike: Second All-out Stoppage Hits NHS 27 April, accessed on 15th July 2025 from https://www.bbc.com/news/health-36145686.
[77] BMA (2016). Call for formal renegotiations. British Medical Association 11 August.
[78] Dyer L, Reeves T (1995). Human resources strategies and firm performance: What do we know and where do we need to go?. International Journal of Human Resource Management 6(3): 656–670.
[79] Cherkaoui M (1977). Bernstein and Durkheim: Two theories of change in educational systems. Harvard Educational Review 47(4): 156–166.
[80] Bourdieu P (1984). Homo Academicus Paris: Editions de Minuit.
[81] Johnson P (2006). This out-of-date business model will erode Oxford's distinctions The Financial Times 13 November.
[82] Smithers R (2005). City academy raises exam grades by paying pupils cash bonuses. The Guardian. 17 November.
[83] Hirschman A (1970). Exit, Voice and Loyalty Cambridge MA: Harvard University Press.
[84] House of Commons (2015). Academies and Free Schools. Fourth Report of the Select Committee on Education Session 2014–15, January.
[85] Rousseau JJ (1757/1961). Émile, Ou de l'Education Paris: Garnier.
[86] Whitehead AN (1932). Essays in Science and Philosophy New York: Philosophical Library.
[87] Becker BE, Huselid MA, Pickus PS, Spratt MF (1997). HR as a source of shareholder value: Research and recommendations. Human Resource Management 36(1): 39–48.
[88] Boxall P, Purcell P (2008). Strategy and Human Resource Management Basingstoke: Palgrave Macmillan.
[89] Bologna Declaration (1999). Joint Declaration of the European Ministers of Education Bologna. 19 June, European Union.
[90] Oliveira TC, Holland S (2008). Retórica e Realidades Nas Reformas do Ensino Superior e No Processo de Bolonha. Ensino Superior, Revista Do SNE Sup 29: 19–33.
[91] Department of Health (2004). National Standards, Local Action. In: Health and Social Care Planning Framework 2005/06–2007/08, London: HMSO.
[92] DHSS (1983). The NHS Management Enquiry The Griffiths Report London: Department of Health and Social Security.
[93] Carvel J, Woodward W (2006). Thousands of jobs go in NHS cash crisis The Guardian 24 March, accessed on 15th July 2025 from https://www.theguardian.com/money/2006/mar/24/publicfinances.health.
[94] Bolton SC (2004). A Simple Matter of Control?. NHS Nurses and New Management. Journal of Management Studies 42(2): 316–333.
[95] Stern S (2006). It is time to end our unproductive fixation with productivity The Financial Times 11 April.
[96] Womack JP, Jones DT (2005). Lean Solutions: How Companies and Customers Can Create Wealth Together New York: Simon Schuster.
[97] Teich ST, Faddoul FF (2013). Lean management – The journey from Toyota to healthcare. Maimonides Medical Journal 4(2): 1–9.
[98] Kaplinsky R, Posthuma A (1994). Easternisation: The Spread of Japanese Manufacturing Techniques to Developing Countries London: Frank Cass.
[99] Brynjolfsson E, MacAfee A (2014). The Second Machine Age: Work, Progress and Prosperity in a Time of Brilliant Technologies NY: Norton.

[100] Frey CB, Osborne MA (2013).The Future of Employment: How Susceptible Are Jobs to Computerisation? Retrieved from http://www.oxfordmartin.ox.ac.uk/downloads/academic, last accessed 2013-09-17.
[101] Lanchester J (2015). The robots are coming. London Review of Books 37(5): 3–8.
[102] Holland S (1993). The European Imperative – Economic and Social Cohesion in the 1990s. Foreword Jacques Delors Nottingham: Spokesman Books.
[103] Croucher S (2014). Lux leaks: Jean-claude juncker facing credibility crisis after latest Luxembourg tax avoidance Scandal International Business Times 6 November, accessed on 15th July 2025 from https://www.ibtimes.co.uk/lux-leaks-jean-claude-juncker-facing-credibility-crisis-after-latest-luxembourg-tax-avoidance-1473543.
[104] Manchin R (2005). After the Referenda Brussels: Gallup Europe.
[105] European Council (2000). Presidency Conclusions Lisbon, June.
[106] European Council (1994). Presidency Conclusions, Essen, December.
[107] Dore R (1986). Flexible Rigidities and Industrial Policy in the Japanese Economy London: The Athlone Press.
[108] Dulebohn JH, Bommer WH, Liden RC, Brouer RL, Ferris GR (2012). A meta-analysis of antecedents and consequences of leader-member exchange. Journal of Management 38(6): 1715–1759.
[109] IMF (2016) Structural Reforms and Productivity Growth in Emerging Market and Developing Economies. Working Paper WP 16/15.
[110] Womack JP, Jones DT, Roos D (1990). The Machine that Changed the World New York: Rawson Associates.
[111] Taylor FW (1911). The Principles of Scientific Management New York: Harper & Brothers.
[112] Purcell J, Hutchinson S (2007). Front-line managers as agents in the HRM-performance causal chain: Theory, analysis and evidence. Human Resource Management Journal 17(1): 3–20.
[113] Guest DE, Bos-Nehles AC (2013). HRM and Performance: The Role of Effective Implementation. In: HRM and Performance: Achievements and Challenges, Guest DE, Paauwe J, Wright P (eds) Chichester: Wiley-Blackwell, 79–96.

Diana Santos Fernandes and Carolina Feliciana Machado

5 Human resource management in the health system: in the never-ending quest for productivity improvement

The Portuguese National Health Service case

Abstract: The need to pay attention to health human resources was underlined by the World Health Organization in its most recent assessment (2009) of the current national health program and performance of the Portuguese National Health Service (NHS) system. This chapter focuses on health human resource management (HHRM) capabilities in the Portuguese NHS to determine whether there is a paradigm of strategic management, concluding that it does not. We argue that there is a "myopia" problem in the standard of HHRM because it focuses on administrative costs and not on the broader workforce, which is able to give it sustainable competitive advantages. This study mostly uses the collection and analysis of secondary quantitative data, presenting the observed performance in human resource management in the Portuguese NHS compared to the expected (strategic HHRM), interpreting the results in a deductive way. We highlight the originality and value of this work as an analytical document and as a script to apply in the transition to a strategic HHRM approach in the Portuguese NHS.

5.1 Introduction

This chapter discusses health human resource management (HHRM) models, presenting as the research problem the analysis of the dynamics in the Portuguese National Health Service (NHS), in order to understand the reason for its inefficient performance. We aim to determine the possibility of the existence of a strategic management paradigm in this public sector; the reason we argue that the improper investment of the Portuguese NHS in its human capital, which can supply a sustainable competitive advantage, is the source of its inefficient performance; this will be the main working hypothesis.

Taking into account these issues, this chapter assumes a great relevance, as in recent evaluations the World Health Organization (WHO) had outlined weaknesses in Portuguese NHS management, focusing in particular on the challenges at the HHRM level, which can threaten its sustainability [1]. We also highlight recent dissatisfaction of users and workers with Portuguese NHS management. Indeed, for workers at a public service focused on human welfare, such aims are undermined by the excessively economic vision of its functions, which is guided by cost savings.

https://doi.org/10.1515/9783112206775-005

This research is divided into three sections. First we will develop a theoretical revision in the strategic human resource management (SHRM) field, focusing on factors that have encouraged the evolution of human resource management (HRM) policies and practices in the direction of this paradigm and its main characteristics, which will be crucial as a framework for this chapter. Having established this support, at the end of the first section we will present the adopted theoretical model as well as the corresponding hypothesis to guide our approach to the topic. The second section will explain the used methodology, focusing on the methods and processes of data collection and analysis. The third section will present the collected data and their analysis, discussing results in a deductive fashion. We will elaborate on a characterization of the HHRM model in the Portuguese NHS, comparing the corresponding observed and expected performances.

As we will see, the Portuguese NHS does not represent a strategic HHRM standard, and we will seek, in describing its actual standard, to show its gaps with a view to adopting the desired standard to achieve a more efficient HHRM. It is in this connection that we talk about the originality and value of this research: it is an analytical document and a guide with suggestions to be applied in the transition to a SHRM in the Portuguese NHS.

5.2 Literature review

5.2.1 Strategic dimension of HRM

HRM consists in the close interdependence between a knowledge pool and an activity to achieve results, influencing the collaborators' behaviors and attitudes through a management system. This comes from the growing awareness under which the organization of human capital can hardly be imitated, contrary to its traditional competitiveness factors, changing the role of workers, who become the strategic basis of the organization. In this way, we justify the relevance of SHRM on the premise, supported by empirical evidence, that organizational performance is influenced by the pool of HRM practices, defined, and executed by the organization, with this assumption serving as the basis of the theory as well as of the research to be presented [2].

In the last decade HRM has become increasingly recognized as an important part of business, as its conceptualization developed from the concepts of costs, administration, training, and collective negotiation to another based on the concepts of investment, management, development, and teamwork. Such a change in paradigm reveals, indirectly, the erosion of traditional competitive factors, showing, directly, the importance of considering human resources (HR) as a competitive and strategic advantage in an organization's survival. HR can act, in the view of Wright and MacMahan [3,

p. 93], as "human capital."[1] The authors propose that it will serve as a mediator between performance and HRM practices once the relation between these last two named issues is established as empirically demonstrated in the literature, for instance, in the work of Combs et al. [4]. Lengnick-Hall and Lengnick-Hall [5], in accordance with the aforementioned authors, clarify the concept of competitive advantage, arguing that it is related to capabilities, resources, relations, and decisions that allow an organization to capitalize on opportunities and avoid competitive threats within the sector in which it operates, which will require the adoption of SHRM.

The contemporary HRM vision arose after the 1990s, as organizations faced a multitude of technological, economic, social, and political-legal challenges and needed to reorganize themselves at the structural and management levels to survive in the markets. The HR function developed and sought answers to current challenges through the reconciliation of economic options, the desire for profits, and HR evolution, which is why, in this context, organizations focus on internal resources, assuming from that point forward that HR would be their first strategic resource. Organizations placed a bet on SHRM, according to Lengnick-Hall et al. [6], that organizations would become more resilient, more competitive in the current international environment, and increasingly complex as a result of globalization.

Developing a critical literature review we verify that this idea is systematized by Wright et al. [7], solidifying organizational theory. We then emphasize how organizations understand their internal resources, which create value for them, leading to sustainable competitive advantages, as systemic imitation requires an understanding of exactly how all its elements interact, which is difficult, rendering infeasible the simple copying of isolated policies of competitors.

Related to this perspective, Ulrich [8] presents another theoretical categorization, proposing an organizational orientation toward key competencies, following the development of decentralized processes of decision-making, since this stimulates workers' greater commitment and motivation. These ideas are systematized by Pfeffer [9], who proposes that HR needs to play a new role in organizations, taking a central part, not in traditional activities, such as hiring, remuneration, or administrative support, but by developing enriching and useful solutions to organizational issues since HR is an "invisible asset" that can bring sustainable competitive advantages. To this end, Huselid and Becker [10, p. 426] suggest that SHRM adoption must be articulated in the

1 The concept of "human capital" has its origins in the economic literature. Becker (1964, as cited in Wright and MacMahan [3]) systematizes this concept as referring to the "knowledge, information, ideas, skills and health of individuals," issues therefore included in an organization (Becker, 2002: 1, cited it [3]). In a later contribution, the authors clarify that "[c]omparing human capital to financial or physical capital, (. . .) all are forms of capital in the sense that they are assets that yield income and other useful outputs over long periods of time. On the other hand, the uniqueness of human capital stems from the fact that people cannot be separated from their knowledge, skills, health or values in the way they can be separated from their financial and physical assets" (Becker, 2008, as cited in [3]).

following steps: operationalize strategy, operationalize strategic capabilities, operationalize strategic jobs, measure workforce differentiation, implement an HR architecture, determine impact of workforce differentiation on employees, integrate qualitative and quantitative data collection and analyses – all this in order to stimulate organizational performance improvement.

We live in a time of large-scale economic, technological, and social changes, where information access, volume, and speed are increasingly difficult to absorb both by individuals and organizations. Thus, the urgency of systemic thinking to understand this more complex environment due to globalization seems all the more pressing. On one hand, systemic thinking is understood as an interesting field of research; on the other, it is also possible to observe challenges that can constraint knowledge evolution in this subject.

Systemic theory, applied to organizational activities, has had a huge impact on SHRM progress. Given the interdependence among HRM policies and practices, it is necessary to focus on its conceptualization as a coordinated system in order to analyze its impact on organizational performance, which requires that the system be greater than the simple sum of its parts: it is only by the interaction of the parts that specific characteristics will emerge, which is why Jamrog and Overholt [11] emphasize that the key to a theoretical approach to HRM lies in the mutual interaction of the different parts of the system.

Thus, research interest in HRM policies and practices systems and their impact on organizational performance has been increasing, on the assumption that the key to organizational success lies in the mutual satisfaction of collaborators and organizations' interests and needs. We are witnessing a new conception of HRM in organizations, assuming a work force highly involved in organizations, highly flexible in their roles and skills, focusing on such key issues as selection, socialization, training, and reward systems development [12].

According to Wright and McMahan [13], SHRM is being undermined by severe critics since without a strong theoretical basis to understand specific SHRM policies and practices, it is difficult to distinguish SHRM from traditional HRM. Nonetheless, the efforts aimed at SHRM theorization are motivated by a recognition of the strategic and symbolic dimensions of the human factor and its impact on organizational efficiency since human capital is conceived as an invisible asset in organizations. This follows Pfeffer's [9] ideas on the dynamic and relevance of organizations' sustainable competitive advantages.

Despite the failure of academic researchers thus far to agree on a definition of SHRM, there is a broad consensus that such policy aims to achieve a match between HR and organizational skills, developing them mutually, taking into account that one will incentivize the other, in a logic of interdependence.

In Watson's (1985, as cited in [14]) perspective, both concepts (HRM and SHRM) distinguish themselves in the value system that they advocate: due to the work force hierarchical control, promote workers' commitment and autonomy in organizational activities, explain why paradigm metamorphosis requires the establishment of a different psychological contract based on reciprocal interactions as well as on high levels of trust and mutual satisfaction. The difference between paradigms is visible in the adoption of opposite HRM policies and practices. Work force planning is undergoing a transition from that of a "police nature" developed by supervisors to one that is participative, proactive, strategic, integrated, and having a long-term dynamic. The self-regulation paradigm contrasts with this by its opposition to external control, as well as individual relationships on a basis of trust and not of intimidation or retaliation. It privileges flexibility and versatility of roles and sees HR as something to invest in, as opposed to a monolithic HRM concerned exclusively with cost cutting.

In accordance with the growing awareness that the social and economic domains are very closely interconnected – a consequence of globalization dynamics – SHRM is legitimized as a concept, as supported by Prahalad and Hamel's [15] reflection that in this model jobs and skills are better integrated, with HR having a dynamic vision of organizational resources. The concept of competence becomes essential for achieving organizational success, as success must happen increasingly in an environment that remains turbulent and is undergoing continuous change: the contemporary globalized environment. Therefore, in the new HRM policies and practices, increasingly of a strategic nature, organizations focus on customization, adaptation, mobilization, sharing, and anticipation, issues that will be discussed in greater depth when we present the theoretical model that supports this chapter [16, p. 34].

Ulrich (1996, as cited in [14, p. 52]) systematizes the two main vectors whereby HRM policies hang together: the *strategic* (long term) and the *operational* (short term), where SHRM activities should embrace individuals, in such processes, as a whole. To articulate even more the degree of systematization of this theme, Tavares and Caetano [14, p. 52] present, referring to Ulrich (1996), the four main aims of SHRM, combining the aforementioned strategic and operational axes: SHRM, transformational and changing management, organizational infrastructure management, and worker contribution management. Thus, SHRM is both a concept and a set of practices, which is why it is so complex.

5.2.2 Theoretical framework definition and research hypothesis

In our understanding, health human resources (HHRs) has a twofold dimension, where capital plays a dual role of individual and agent of health care practices. A health system is grounded in three different types of capital – structural (health organizations), professional (HHR), and citizens (users) – and it is from their interaction that knowledge is created, based on the structural capital system and shared information and on citizens and the organizational professionals' existing knowledge. This results in an increase in organizational intellectual capital, which requires coordinated action by all actors in management policies and practices [17].

To achieve the goals proposed in this chapter, we have elaborated a theoretical model considering the inputs given by three different authors, which is represented in Figure 5.1.

This was the option that resulted from some gaps found in an initial theoretical revision that we made on this topic. Indeed, it was not possible to find in the literature HRM models of analysis specific to the health care sector, as this thematic is, in this knowledge field, very recent. It was only possible to find general standards that must govern SHRM, not as applied to a specific field of activity, as well as specific characteristics that management in the healthcare sector obey, applied, however, to some particular national contexts. Moreover, and according to the preceding literature review, we also observed that the debate about SHRM remains, which is why, to overcome these constraints, it was necessary to develop a specific theoretical model adapted to the main aims of this study.

In what follows, we present the adopted theoretical model and its dynamic in what concerns the analysis of the problematic examined in this chapter. We adapt Dussault and Souza's [17] theoretical model on the logic of the analysis of the situation with HHR because it provides analytical categories for understanding HR dynamics in a health unit, in which health goals are attained in a particular moment and in a specific context. To this end, the authors highlight three main processes at the HHRM level.

Competencies development refers to the knowledge production process and the competencies necessary to "achieve the levels of performance required by the health services goals" [17, pp. 13–14] in each professional category. *Worker distribution* focuses on the professional allocation process in different services, regions, and subregions of the country, ensuring equality of access to all services for the entire population. About *personnel management* it looks to "guarantee an appropriate work environment and an adequate level of performance" of HHR [17, p. 23].

At the base of health sector HRM policies and practices (applying the development of a theoretical model to the specific context of analysis in this work, the Portuguese NHS) lie many relevant inputs that constrain them, namely, external (e.g., health episodes at the international level that have repercussions in the national reality; governmental policies that affect NHS regulation – financial, recruitment, rewards) and internal (e.g., ministry of health policies, already established power relations in the NHS and HHR). These inputs will have repercussions in HRM policies and practices in the NHS at the distribution, management, and competency development level of the respective HHR. Thus, "all decisions, actions and inactions that affect production, distribution and personnel management are policies that need to be taken into consideration given the HHR analysis of situation" [17, p. 14]. Although this issue is considered in Dussault and Souza's model, we have decided to develop and clarify it, adapting it to our specific case study.

The coordination of all participants is required in the planning and execution of policies and practices in the health sector, as each decision dimension is, according to

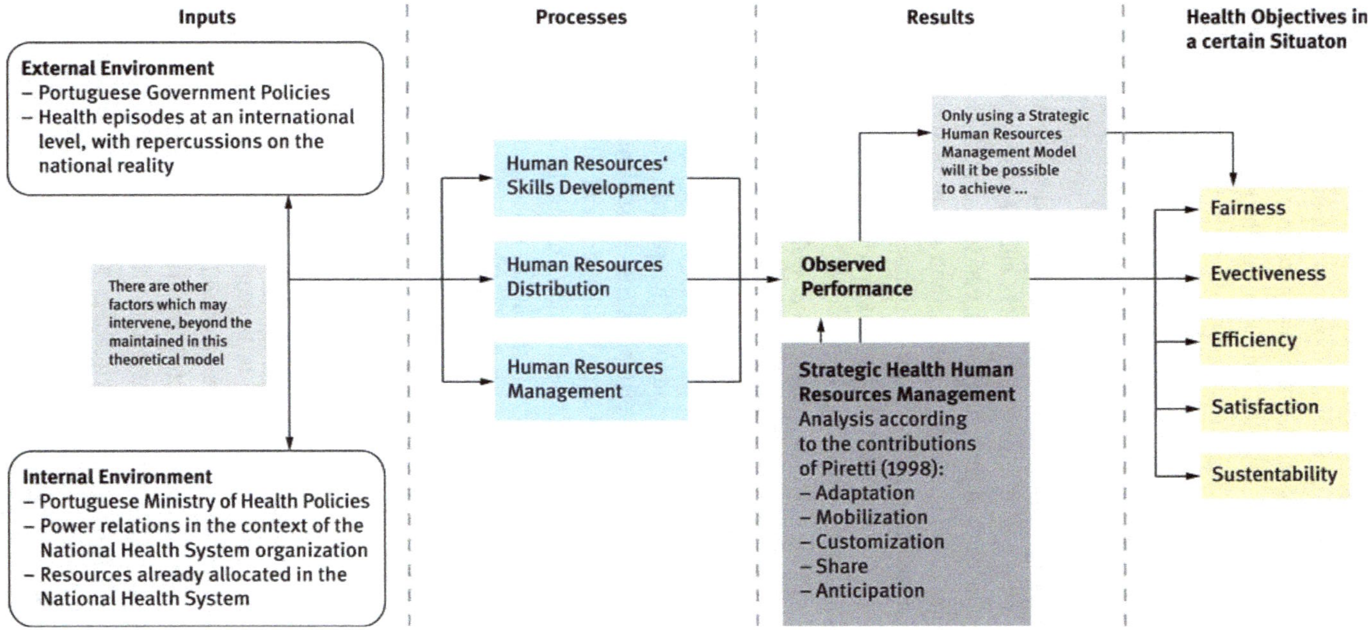

Figure 5.1: Scheme of adopted theoretical model. The logic of the situational analysis of HR in the health sector, particularly in the Portuguese National Health System (NHS), taking into account its management according to a strategic paradigm (source: adapted from [16, p. 54; 17, p. 13]).

its nature, the responsibility of individual actors. Then the two previously mentioned inputs are correlated since the resulting external dynamics and orientations will be configured around the internal environment, providing more or less favorable conditions for the further implementation of HHRM. This is one of the changes (or better, clarifications) that we have inserted in the theoretical model of the cited authors.

Noting that such a model will be incomplete in terms of the present research, considering the hypothesis that we put forth (see end of this section), we have decided to introduce into the basic theoretical model the contributions of Peretti [16, p. 34], which will provide the key vectors by which an organization can maximize its performance, focusing on SHRM. By *adaptation* the organization will address internal and external requests through the use of different types of flexibility. *Mobilization* applies to the involvement of all hierarchies in HR commitment, which goes through information generation and dissemination, opening up options and means of solving problems. In turn, *customization* embraces global rewards, training, and information; career, competencies, and time management, reconciling workers' expectations with organizational constraints. *Sharing* ensures that all managers take part in HRM in collaboration with the director of the corresponding job that will promote the new concept of a shared job. By *anticipation* we reinforce the need for forward-looking job management in the medium and long terms, taking into consideration environmental unpredictability.

Explaining generally the dynamic of the adopted theoretical model, we verify that Peretti vectors that lead to SHRM in the Portuguese NHS must be applied in each of the processes of Dussault and Souza in the situational analysis of HHR in this service. Such processes will require a set of inputs, influencers of the dynamics that we can verify in each of them. Only by this logic will we attain efficient organizational performance in the Portuguese NHS, therefore we will be able to detect if in the Portuguese NHS is presently applied a SHRM model, appearing from this conclusion the reasons that underlie to its inefficient performance.

Taking this into account, we have established a research hypothesis whose theoretical fundamentals lie particularly in Huselid et al.'s [2] reflections. Based on the heterogeneity premise among organizations on what concerns human capital, competitive advantage is possible only if an organization guarantees that the added value that is possible as a result of activities performed by its workforce is seen as a unique resource, difficult to imitate and replace, as we emphasized in the theoretical review section. In this vein, SHRM practices embrace activities through which the organization generates human capital able to achieve these conditions and in this way promote positively organizational performance. Then, SHRM activities will serve as the means to achieve efficiency because they will help to ensure that the organization's HR will not be easily imitated due to the complexities of culture and social dynamics and causal ambiguity inherent in SHRM practices, such as empowerment, schemes based on team collaboration, and long-term talent development, among other factors.

Thus, such arguments suggest the following mutually exclusive hypotheses:

H_0: *The Portuguese NHS presents a standard of SHRM that demonstrates that its performance, in terms of its HHR labor, will be efficient.*

H_1: *The Portuguese NHS does not present a standard of SHRM that demonstrates that its performance, in terms of its HHR labor, will be inefficient.*

The observed performance in the Portuguese NHS, in terms of the labor performance of its HHR, is the dependent variable, and the existence and application of an SHRM model will be the independent variable.

5.3 Methodological considerations

This study articulates descriptive research, aiming to determine the HRM standard in the Portuguese NHS, presenting phenomena as they are in reality. In this case, it is necessary to collect a data set so that, based on a theoretical model that provides tools related to the dynamic of an SHRM model, it becomes possible to identify variables and inventory facts that make it possible to conclude whether or not the HHRM standard in the Portuguese NHS is strategic. Thus, this research follows deductive logic as it develops from a conceptual and theoretical structure that is later tested based on empirical observations.

The adopted research process emphasizes a quantitative feature since it is mainly concerned with collecting statistical data and assembling and analyzing numerical data, although other data of a qualitative nature may also be relevant to our study. The research adopts a positivistic paradigm, starting from the assumption that research develops in a logical direction, deducing that an act of researching reality has no effect on that reality, as the researcher seeks facts and the corresponding causes of social phenomena, placing little importance on the subjective position of the person.

In terms of the results, this study's focus is basic research, as it looks to contribute, generally, to the scholarly knowledge base in the field of HRM modeling, with a special emphasis on strategic models, in the health field. However, this study could also serve to stimulate, in future research, the development of objective measures that would act as a guide in the implementation of SHRM in the Portuguese NHS or in other similar contexts. In this vein, this study could, in future research, see further development with a more applied character. We highlight its originality and value as it is assumed not only as an analytical document but also as a guide to be applied to behavioral changes, bearing in mind SHRM in the Portuguese NHS.

The data to be collected are largely quantitative and secondary in nature and focus on the statistics presented by the Portuguese Statistical Institute (INE). Other sources include research reports and surveys that make observations on issues specific to HHR as elaborated by the Portuguese Medical Association or the Portuguese Nurses Association. We have also used official information released by the Ministry of

Health, projections elaborated by the Portuguese Central Authority of Health Services (ACSS), as well as Portuguese legal documents. Contributions from scientific books and journals and other studies addressing this theme were also used. This was necessitated by the dearth of information on the subject available for study and by the difficulties in accessing the information, as well as the need to collect the maximum amount of information about the past and current situation of the Portuguese NHS. This will be a limitation of the present study, which is why the methods used in data collection were based on the analysis of documents and bibliographical filtering since primary data could not be gathered and other data collection mechanisms that would have allowed us to conduct a more in-depth analysis of the data at our disposal were not available.

Data was collected in the 2000s and 2010s because it is important to use data from a reasonably long but also relatively recent period of time. In this way, the subsequent data analysis will allow us to extrapolate, in a more reliable way, evolving trends in the Portuguese NHS, in terms of its HRM. Thus, this research represents a longitudinal study.

With such data, we intended to collect information about the performance of HHR in the Portuguese NHS, in each of the processes of the adopted theoretical model, in an attempt to determine its present HRM model (which we call "observed performance").

Indeed, to analyze the process of *HHR competency development*, we undertook the collection of evidence relative to Portuguese HHR training, in part relative to all newly trained HHR, to the content and length of training, to the constraints on doctors and nurses regarding the formal launching of their careers, and to the accreditation agencies of the education establishment. To analyze the process of *HHR distribution*, we gathered evidence on the surplus or shortage of health professionals as well as of HHR personnel based on nationality and region of the country where they work. Also important was the amount of HHR allocated to provide primary and secondary health care and relative to the share of HHR based on clinical or professional category, and further the amount relative to provider-to-patient ratio by type of service. To analyze the *HHR management* process, we analyzed the criteria used in worker allocation, working hours, legal documents in Portuguese NHS management, HHR recruitment methods in public services, and the (non-)existence of incentives for improving distribution and staff retention.

The analysis and further interpretation of data were processed using deductive logic (supported by the assumptions of the adopted theoretical model and the information mentioned in the earlier literature review), through content analysis, looking to elaborate on an actual performance description of the Portuguese NHS, occasionally also using a comparative methodology. This strikes us as an appropriate approach since we base our analysis on a longitudinal study in terms of a time period comparison to extrapolate the behavior standards of the involved variables, defining them, describing their behavior, and outlining their evolutionary dynamic.

Yet, based on the analysis and corresponding descriptions of HHR performance in the Portuguese NHS, we will reveal the observed performance (present) in its management. Articulating the presented data analysis methodology, we will verify whether that performance matches the expected performance, which will focus on SHRM. In the present study it will not be possible to analyze each of the aims in the health sector. Thus, our approach consists in emphasizing the comparison of the observed and expected performances with respect to HRM in the Portuguese NHS. In addition, we argue that efficiency will be possible only in a certain context and at a given time if SHRM is adopted.

5.4 Data presentation and analysis and discussion of results

5.4.1 Processes in the logic of HHR situation analysis in Portuguese NHS

In what concerns HHR competency development, we have verified that continuous professional development covers all health professions, answering to the educational needs of different elements of a multidisciplinary work team within the quality improvement scope of the health unit concerned (Pringle, 2000, as cited in [18]). However, the transition to this paradigm never happened in the Portuguese NHS [19].

Most educational institutions that develop training in the health care field are accredited. However, a significant number of such institutions are only in the preliminary stages of accreditation. There are extensive training requirements and a series of professional self-regulation procedures by multiple entities that require evaluations between different professional degrees that do not belong in programs structured and articulated by local training departments, much less entail improvements in the health units' quality. The approval to open up courses in the health field, in the public or private sector, as well as the establishment of a limited number to accede to the university, or in available jobs to medical or nursing specializations in the country, is limited by governmental directives, which limit the existence of a higher workforce in the Portuguese NHS (Rodrigues, 2002, as cited in [18, p. 286]).

Between 2000 and 2010, admissions to training programs for doctors, nurses, dentists, and pharmacists saw continual growth. The number of trained specialist doctors has clearly grown, at a rate exceeding 120% between 2005 and 2010. The annual average growth of university graduates between 2000 and 2008 is positive in these professional categories, except for pharmacists in 2004, which saw a decrease of 41% [20, pp. 23–25].

With regard to **HHR distribution**, since 2002 we have witnessed continual growth in the number of doctors, nurses, pharmacists, and dentists. The ratio of nurses to doctors from 2002 to 2,010 remained constant at 1.2 [20, p. 23]. Thus, we see a "real contrast" between the increase in the supply of doctors and nurses, and its recruitment (demand) by the NHS, although there might be "nuances" to this interpretation, as detailed in what follows (Figure 5.2).

We highlight the negative evolution of nurse recruitment; however, with respect to doctors, the increase is a result of the enrollment of young people in training programs and with short-term contracts, being still necessary to refer to certain clinical expertise where Portuguese NHS shows a deficit in qualified HHR.

We understand this growth as unreal, and it coincides with the exit of doctors from the NHS (due to, for example, retirement, or emigration), which is why the perspective of "quality gains" in terms of assistance and training capacity is, indeed, negative (Figures 5.3 and 5.4 and Table 5.1) [22, p. 29].

The percentage of NHS doctors and nurses who work on basic health care is half that who work in hospitals or in other health-related services, based on data from 2000 to 2007; the number of other health professionals in this field increased during the same period [20].

HHR assignments are based on two criteria: the greater distance of the three more desired geographical areas and the population density of the geographical areas covered. However, these criteria appear to be obsolete, as "they have been contributing to the regional differences of assignment," given the covered population [19, pp. 4–5]. Confirming the situation, we see that the coverage ratio of NHS doctors for every 10,000 inhabitants ranges from 26 in the Lisbon Tagus Valley and central region and 18 in the Alentejo region (Table 5.2 and Figure 5.5) [22].

To overcome the shortage of specialized (expert) health professionals, particularly in certain clinical specialties (areas of expertise), the Ministry of Health imports equivalent foreign HHR as an easy way to solve this problem (Figure 5.6). Nonetheless, Portuguese nationals predominate in the NHS, representing more than 93% of the total, an increase over the 2009 level [22].

With respect to **HHR management**, the importance of the link "Indefinite Labor Contract in Public Functions" diminished between 2005 and 2010 in terms of the evolution of the forms of contractual ties predominant in the Portuguese NHS, reflecting legislative changes in public administration recruitment (Figures 5.5 and 5.7). In addition, the number of people working between 35 and 42 h also dropped, while the number of those working 40 h increased [22].

According to the same source, we verify that the NHS covers the country offering its services according to the two major health subregions, in line with the major national metropolitan areas (Oporto and Lisbon), with more than 40 healthy centers, organized according to geographical area. These integrate more than 100 health extensions, with a total equivalent to thousands of HHR depending on ministry directives. The management of HHR in the Portuguese NHS is vertically integrated and, hence, largely dependent on those in power at a given time. New laws governing hospital management and the network of primary health care centers underline the con-

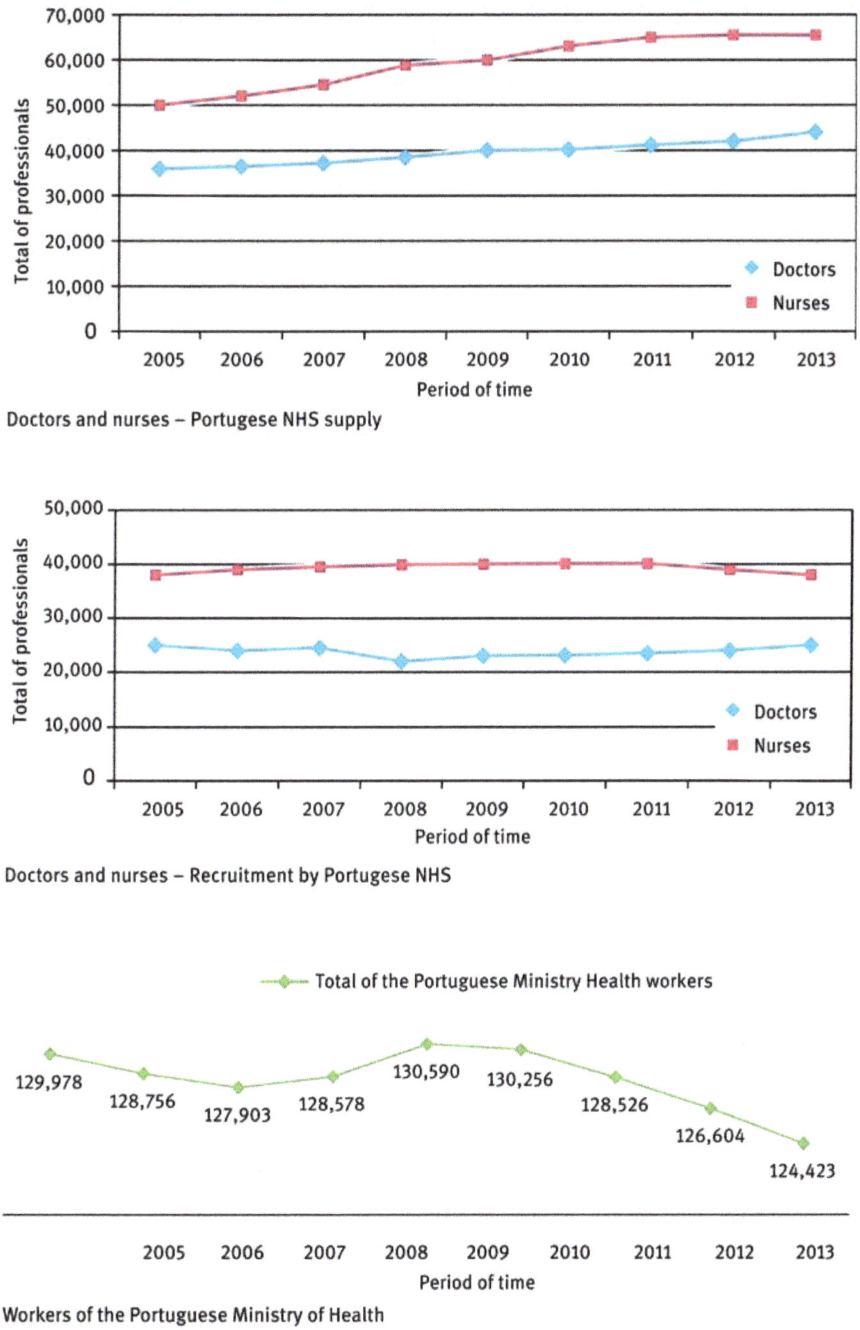

Figure 5.2: Relation between specialized HHR supply and demand in the Portuguese NHS, namely, in doctor and nurse workforces (source: adapted from [21]).

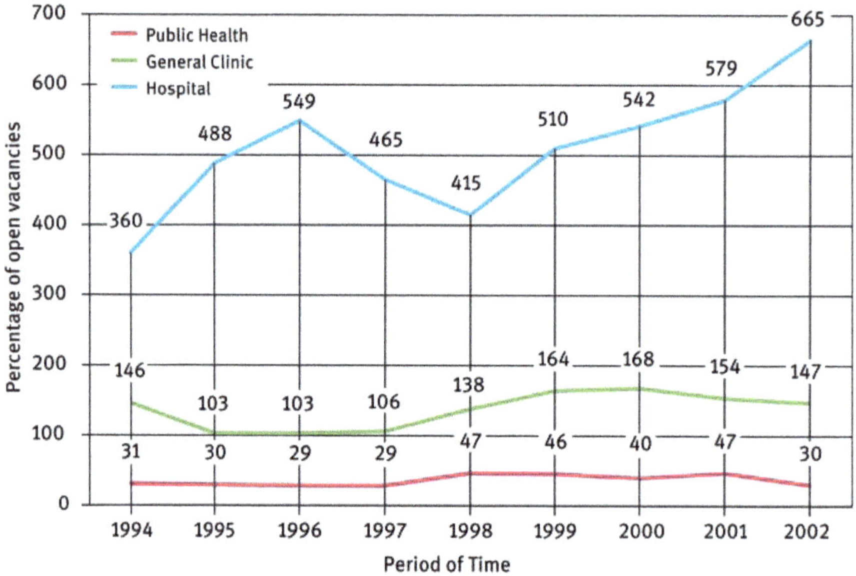

Percentage of open vacancies in the entrance exams of the hospital doctors

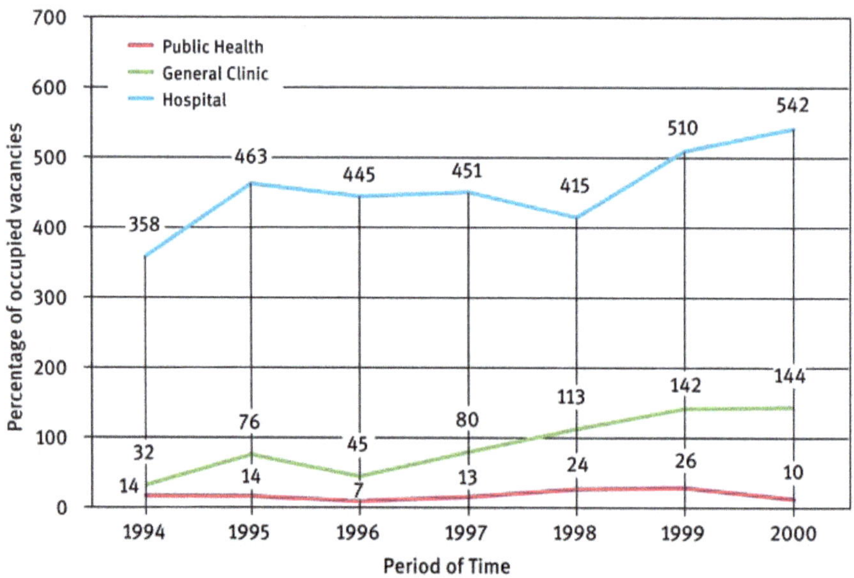

Percentage of occupied vacancies in the entrance exams of the hospital doctors

Figure 5.3: Comparison between percentages of open and occupied vacancies in entrance exams of hospital doctors (source: adapted from contributions of Mission Group participant in elaboration of resolution of Ministry Council 140/98 of 4 December, relative to strategic plan to training in health areas; in OPSS [23]).

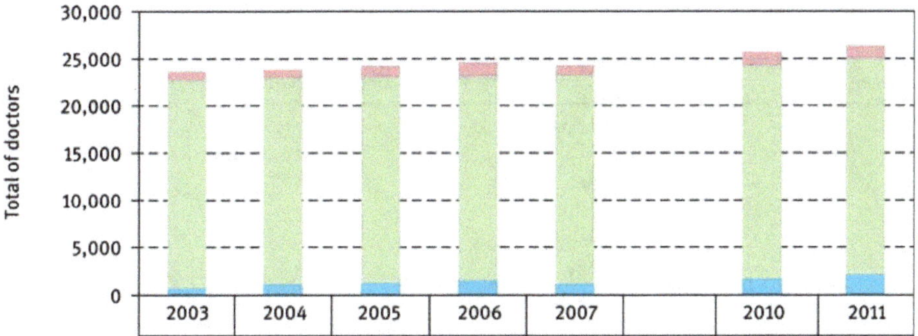

	2003	2004	2005	2006	2007		2010	2011
Exits	929	945	1,217	1,535	1,049		1,438	1,412
Maintenance	21,921	21,765	21,729	21,533	22,068		22,488	22,805
Entrance	789	1,181	1,339	1,584	1,197		1,729	2,190

Medical turnover in the Portuguese HNS

Figure 5.4: Medical turnover in Portuguese NHS (comparison 2002–2011) (source: adapted from [22]).

Table 5.1: Growth rate in total number of clinics, according to clinical expertise, in Portuguese NHS (comparison 2002–2011).

Career	Specialty	Growth (2002/2011)
Hospital		8%
	Pathological anatomy	−18%
	Anesthesiology	26%
	Cardiology	11%
	Pediatric cardiology	16%
	Cardiothoracic surgery	− 1%
	General surgery	12%
	Maxillofacial surgery	7%
	Pediatric surgery	22%
	Plastic surgery	− 1%
	Vascular surgery	26%
	Dermatology	−10%
	Endocrinology	0%
	Estomatology	−19%
	Gastroenterology	5%
	Medical genetics	50%
	Gynecology/obstetrics	−11%
	Clinical hematology	4%
	Immunoallergology	54%
	Immunohemotherapy	− 3%
	Infectiology	54%
	Physical medicine/rehabilitation	3%
	Internal medicine	26%

Table 5.1 (continued)

Career	Specialty	Growth (2002/2011)
	Nuclear medicine	6%
	Nephrology	40%
	Neurosurgery	4%
	Neurology	19%
	Neurocardiology	56%
	Ophthalmology	− 4%
	Medical oncology	231%
	Orthopedics	− 3%
	Otolaryngology	− 1%
	Clinical pathology	−12%
	Medical pediatrics	9%
	Pedopsychiatry	34%
	Pneumology	15%
	Psychiatry	−18%
	Radiology	7%
	Radiotherapy	−19%
	Rheumatology	63%
	Urology	5%
General clinic	General/family medicine	−14%
	No specialty – general clinic	Data not available
Public health	Public health	−24%
Hospital doctor		36%
Other		Data not available
Atypical categories		Data not available
Total general		9%

Source: Adapted from [22].

Table 5.2: Inhabitant ratio by specialist (expert) doctors in NHS, according to respective career and specialty (comparison 2002–2011).

Practice setting	Specialty	Number of inhabitants per specialist (2002)	Number of inhabitants per specialist (2011)
Hospital		914	854
	Pathological anatomy	55,152	67,886
	Anesthesiology	10,675	8,471
	Cardiology	25,455	23,150
	Pediatric cardiology	320,240	279,086

Table 5.2 (continued)

Practice setting	Specialty	Number of inhabitants per specialist (2002)	Number of inhabitants per specialist (2011)
	Cardiothoracic surgery	139,823	143,530
	General surgery	11,307	10,263
	Maxillofacial surgery	330,915	313,971
	Pediatric surgery	171,163	141,508
	Plastic surgery	102,345	104,657
	Vascular surgery	171,881	110,408
	Dermatology	58,742	66,099
	Endocrinology	87,853	88,912
	Estomatology	61,661	76,695
	Gastroenterology	40,686	39,400
	Medical genetics	1,654,574	1,116,343
	Gynecology/obstetrics	4,935	5,731
	Clinical hematology	89,436	87,366
	Immunoallergology	198,549	130,482
	Immunohemotherapy	60,533	63,189
	Infectiology	145,902	95,687
	Physical medicine/ rehabilitation	42,791	42,038
	Internal medicine	8,840	7,011
	Nuclear medicine	320,240	304,457
	Nephrology	71,938	52,057
	Neurosurgery	85,581	83,034
	Neurology	44,319	37,771
	Neurocardiology	157,578	102,521
	Ophthalmology	23,195	24,445
	Medical oncology	310,233	94,784
	Orthopedics	17,146	17,846
	Otolaryngology	34,833	35,502
	Clinical pathology	23,922	27,451
	Medical pediatrics	2,105	1,905
	Pedopsychiatry	155,116	116,827
	Pneumology	30,735	26,936
	Psychiatry	19,504	23,979
	Radiology	27,273	25,762
	Radiotherapy	137,881	173,226
	Rheumatology	230,871	143,530
	Urology	49,637	47,617
General clinic	General clinic – general and family medicine	1,501	1,767
Public health	Public health	22,460	30,081

Source: Adapted from [22].

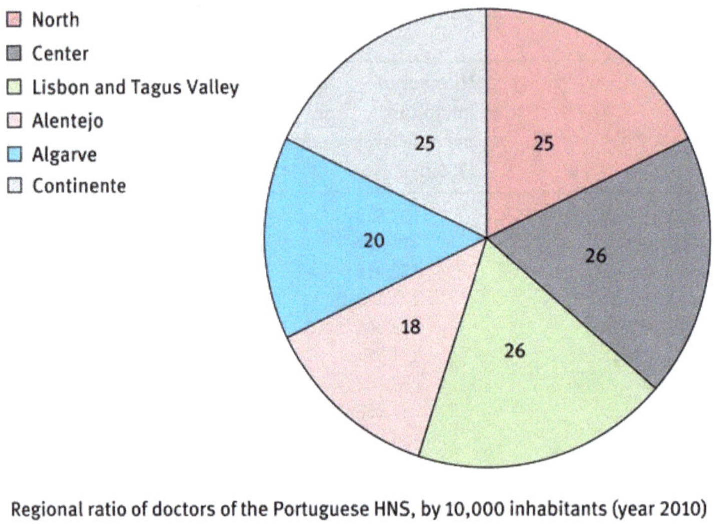

Figure 5.5: Regional ratio of doctors in NHS, by 10,000 inhabitants (2010) (source: adapted from [22]).

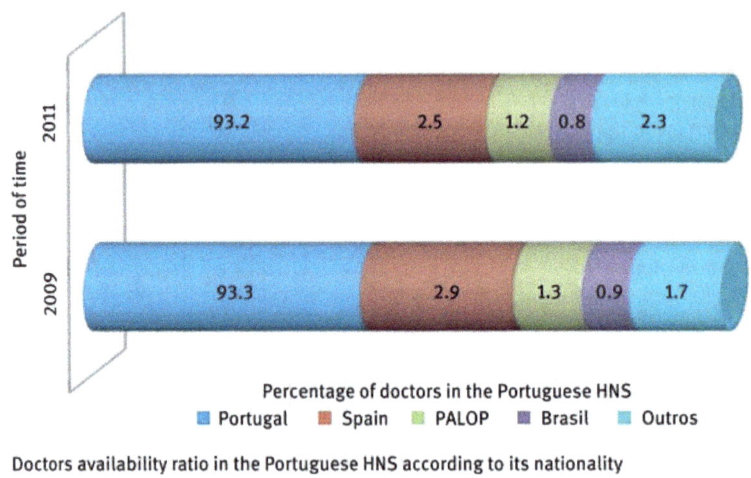

Figure 5.6: Doctor availability ratio in Portuguese NHS by nationality (comparison 2009–2011) (source: adapted from [22]).

straints that this situation could impose on the implementation of SHRM in the Portuguese NHS since they reflect the Ministry of Health's monopolist role in the appointments of hospital and health center directors. This idea relates to that of the last paragraph, showing and emphasizing the Ministry of Health's primary role in the management of the NHS's activities, which is also reflected in the approval of hospitals' activities, which are developed by the ministry, even though such activities could

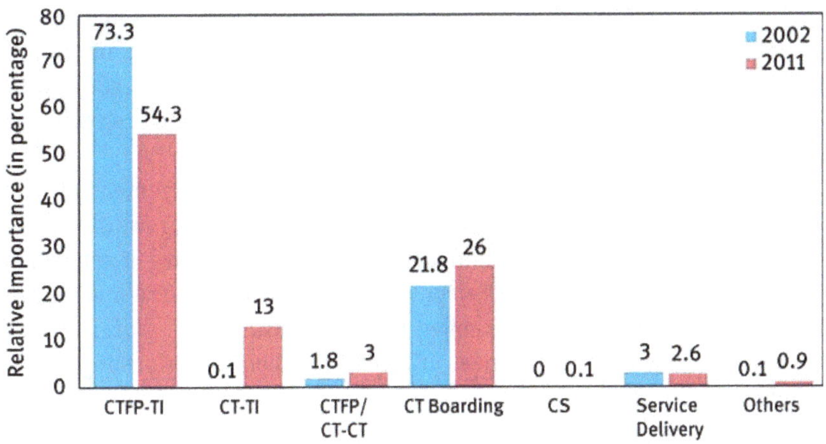

Relative importance of doctors' contractual tie in the Portuguese HNS

Figure 5.7: Relative importance of doctors' contractual ties in Portuguese NHS (comparison 2002–2011) (source: adapted from [22]). CTFP-TI, labor contract in public functions by undetermined term; CT-TI, labor contract by undetermined term (labor code); CTFP/CT-CT, labor contract in public functions and labor code at defined resolutive term; CS, Service Commission.

be overseen by the various health regions [24]. This centralization makes impossible and is contrary to the structure of SHRM in the Portuguese NHS.

5.4.2 Observed performance in HRM in Portuguese NHS compared with expected performance (SHRM), according to five Peretti vectors

With the collected data, we analyzed HR functions to draw conclusions about the performance of HHR in the Portuguese NHS in the period 2000–2010, at the level of the three work processes to achieve health aims, according to the adopted theoretical model and the aims established in this study. Because we believe that it will be possible for the Portuguese NHS to improve its organizational performance only when it adopts an SHRM model, we tried to verify, in the previously given description of such processes, the possibility of its orientation according to the five vectors of Peretti [16].

With regard to the process of **competency development**, we have observed a lack of SHRM in the Portuguese NHS, so continuous professional training presents constraints, in view of its quality and infrequency, as well as its inadequacy for the labor market and health needs. Looking at the particular case of migrant professionals, regulation on the recognition of foreigners' qualifications is weak, and there are several mechanisms: faculties, a medical association (Order), the Ministry of Health, among others.

Considering the **HHR distribution** process, we highlight the actual emphasis in SHRM in the Portuguese NHS, as in the country one can observe several imbalances in HHR distribution, by type of institution, region, or occupational category, which translates into inadequate usage of scarce and expensive resources, limiting the population's access to the corresponding service they need.

In the Portuguese NHS reality, with respect to the **HHR management** process, oversight of workers' administrative tasks – recruitment, placement, reward management, license and pension administration, discipline, and promotion – is centralized at the Ministry of Health and only implemented by personnel units. The involvement of HHR units in the NHS dynamics is limited mainly to bureaucratic processing and, thus, they merely carry out the directives of higher-level bodies, performing routine clerical duties.

The working conditions of HHR in this public service still face constraints in connection with satisfying their own needs and objectives owing to job insecurity, lack of career plans, compensation, and infrastructure problems, which as a result compromise the success of health policies and practices.

Thus, given the description of the observed performance in HRM in the Portuguese NHS, we confirm Hypothesis 1 (H_1) over H_0 since the two hypotheses examined in this study are mutually exclusive. In this regard, we again assert the accuracy of the theoretical model developed to analyze the proposed problematic for this study since the empirical evidence confirms the theoretical postulates: NHS cannot achieve organizational efficiency, in terms of its HHR labor performance, because it cannot accommodate an SHRM model, resulting in certain weaknesses with respect to its workforce, which represents a threat to the long-term viability of this public service. Some of these issues were also addressed by the WHO in an evaluation report of the Portuguese National Health Plan, focusing on the situation of HHR in the Portuguese NHS [1].

To confirm such a conclusion, we explain in what follows why it is not possible to establish SHRM in the Portuguese NHS, supporting the conclusions of Dussault and Souza [20] in terms of the three processes mentioned earlier, taking into account the mistrust by the government, focusing on Peretti [16] vectors. This can be seen as an impulse to the competent authorities design and implement strategic HRM policies and practices in this service, which could be guided by the need to overcome the weak performance observed in each vector.

Regarding **adaptation**, awareness of NHS-specific issues is essential to implementing effective and strategic HRM policies and practices. Its structural renovation will not be enough to undertake effective revitalization initiatives. This will only be achieved by adjusting HHR skills in the organizational strategy. The particular features of the NHS are at the root of its problems, in particular, with respect to "democracy and autonomy," as the main mechanism of coordination (skills standardization, delegated in a central way according the directives of the Ministry of Health) will not be enough to meet all coordination needs either from professionals among themselves or from them in the support sector. In terms of results, concerning this vector, the reflection of Lengnick-Hall et al. [6] is pertinent, as we have established that the Portuguese NHS will be able to achieve resilience only by instituting an SHRM paradigm that will enable it to be more competitive in the current complex international environment, a result of globalization.

Regarding **mobilization**, the Portuguese NHS generally had no clear targets or instructions about HR performance appraisal and was unable to develop, in a systematic way, the respective health services. The careers of doctors and nurses are the only feature that has guaranteed, until now, mechanisms that aim at the evolution of their professionals toward high standards of accountability and compensation, which acts as a control mechanism by the Ministry of Health, as they permit a degree of hierarchy among such professionals [25]. We see in the NHS the proliferation of individual employment contracts, bringing disruption to labor perspectives and job insecurity, which erode values in this public service [25]. In this discussion, we base our interpretation on Huselid et al.'s [2, p. 174] reflections, which address the existence of certain business-related capacities, in addition to techniques that foster the adoption of an SHRM paradigm because of the perception that technical capacities are not enough to ensure an effective HRM standard. Such business-related capacities would allow workers to understand how unique considerations relative to their organization's activities could create in itself HRM-specific needs.

Regarding **customization**, the monthly salary remains the base of HHR compensation in the Portuguese NHS; however, overtime, which are customary and regular, are also a significant part of compensation. According to SHRM, beyond the monthly salary other incentives should be included, like access to or support for training, leaves of absence, flextime, or sabbaticals. These are under discussion in the government,[2] but a lack of knowledge about how to formally appraise the current situation and any kind of cohesive and coherent reflection among professionals and their representatives regarding performance measurement in the different activity areas makes its implementation difficult. In this discussion we follow Huselid et al. [2] in suggesting the need for specific activities in the implementation of a competitive strategy in organizations and in the achievement of their operational objectives. However, there exists a poor understanding about how to reach its effective implementation, having few guides, not existing, a customization in such activities. Therefore, we expect that such HRM strategic activities are relatively few, as is attested by the actual performance characteristics of the Portuguese NHS.

With regard to **sharing**, the Ministry of Health has monopoly control over NHS strategies, which ensure that HHR act simply as its executors. In contrast, in an SHRM

2 To test non-traditional forms of compensation, in 1998 a payment system with a financial component and non-financial incentives was adopted, being necessary particular conditions to the job application [26]. This led to better performance of the target groups despite the lack of official confirmation by the Ministry of Health [27].

paradigm, HRM strategy establishment and execution should be formulated in cooperation with workers in a dynamic of reciprocal interaction and collaboration among the parties. Only in this way will the HHR of the Portuguese NHS be able to act as agents of continuous change. There is a problem with myopia, as the organizational vision of the Portuguese NHS is not shared among all its components. Theoretical grounds that we have used in the discussion of the results achieved in the interpretation of the vectors mobilization and customization are also relevant for this vector interpretation.

Analyzing the **anticipation** vector, we have observed that many entities collect information about HHR in Portugal, according to different criteria. To make things worse, NHS data are not always published in a timely manner, making it difficult to share and access information for a coherent and integrated data analysis. There is also a lack of confidence in the intellectual capital of the Portuguese NHS, as the evolution of this public service is mainly the result of normative initiatives, of a "top-down" logic in terms of power, with little participation from civil society and the HHR who operate the system. Thus, it is difficult to map solutions to anticipate and respond efficiently to challenges that might arise from changes in circumstances in the health environment. On theoretical grounds, the discussion on this Peretti vector is similar to that of the adaptation vector.

5.5 Conclusion

There is not a clear HHR policy or an HRM strategy in the Portuguese NHS that would justify the inefficient organizational performance in this public sector, corroborating previous studies [18]. NHS evolution aims to preserve a public service culture clearly differentiated and qualified, but the maintenance and elevation of this standard require investing in HHR. SHRM has emerged as a bold answer, as it is based on another type of psychological and social contract, valuing human capital, and new forms of relationships with workers.

HRM practice is not neutral as it reflects the values of one organization and its search for objectives related to the adopted values. We argue that the problem that makes impossible SHRM in the Portuguese NHS lies in the myopia of current policies, mainly those of the Ministry of Health, the system's main overseer, because it focuses on an administrative and bureaucratic standard, motivated solely by cost savings.

The conclusions drawn here align with findings in the literature on emergent tendencies that focus on strategy, which is why we share Becker and Huselid's [12] point of view, as we argue that the current theory should expand to focus on the effective implementation of a strategy in the key variables of mediation between HR architecture and achieving organizational performance, and in this way, the organization will reach a level of differentiation, improving its performance.

This will be one of the more critical dimensions of analysis in the SHRM sphere once it looks to determine the best way to evolve so as to improve its operational efficiency to bring about the transformation of its practices and establish sustainable competitive advantages for the organization.

This work outlines directions for future research that will be critical for developing mediation tools. Thus, based on the study presented in this chapter, it is vital to stimulate more systematic work on the relationship between quality standards for individual work units and the organization as a whole. It is therefore of the utmost importance to stress a combination of three axes: performance practices in use in a unit, the identification of areas of improvement should be made, and the individual and team education needs. In this way, it will be possible to attain team commitment, at the individual and collective levels, as HHR in the Portuguese NHS will recognize their contribution to improving services. This is the main recommendation that this study makes for the development of SHRM in the Portuguese NHS, which is why legal mechanisms for its implementation and monitoring, and HHR performance appraisal models, will be topics of future research, on the basis of the various gaps presented in each of Peretti's vectors, as indicated in our description of the observed performance in the Portuguese NHS.

This work also presents limitations, primarily owing to a scarcity of information on this topic and to the difficulty of accessing the available information (a result of the fact that this information is usually collected and interpreted by many entities, according to different criteria, which makes it difficult to analyze). Indeed, it was not always possible to meet all of our research goals with respect to the chosen period of analysis, which made it impracticable to make comparisons and formulate standards on the evolution, in the same circumstances, of certain factors. Thus, the topics covered in this study may be investigated in greater depth in future research.

References

[1] Organização Mundial de Saúde, Departamento regional para a Europa (2009). Evaluation of national health plan of Portugal (2004–2010). Copenhaga: Publicações da OMS. Accessed in 10 October 2015, at http://www.acs.min-saude.pt/files/2010/03/WHO-Evaluation-NHP-Final-March16.pdf.

[2] Huselid M, Jackson S, Schuler R (1997). Technical and strategic human resource management effectiveness as determinants of firm performance. The Academy of Management Journal 40(1): 171–188.

[3] Wright P, McMahan G (2011). Exploring human capital: Putting human back into strategic human resource management. Human Resource Management Journal 21(2): 93–104.

[4] Combs JG, Liu Y, Hall AT, Ketchen DJ (2006). Do high performance work practices matter? A meta-analysis of their effects on organizational performance. Personnel Psychology 59(3): 501–528.

[5] Lengnick-Hall C, Lengnick-Hall M (1988). Strategic human resources management: A review of the literature and a proposed typology. The Academy of Management Review 13(3): 454–470.

[6] Lengnick-Hall C, Beck T, Lengnick-Hall M (2011). Developing a capacity for organizational resilience through strategic human resource management. Human Resource Management Review 21: 243–255.

[7] Wright P, Dunford B, Snell S (2001). Human resource and the resource based view of the firm. Journal of Management 27(1): 701–721.

[8] Ulrich D (1998). A new mandate for human resources. Harvard Business Review January–February 76(1): 124–135.

[9] Pfeffer J (1994). Competitve Advantage through People Boston: Harvard Business School Press.

[10] Huselid M, Becker B (2011). Bridging micro and macro domains: Workforce differentiation and strategic. Journal of Management 37(2): 421–428.

[11] Jamrog J, Overholt M (2004). Building a strategic HR function: Continuing the evolution. Human Resource Planning 27(1): 51–62.

[12] Huselid M, Becker B (2006). Strategic human resources management: Where do we go from here?. Journal of Management 32(6): 898–925.

[13] Wright P, McMahan G (1992). Theoretical perspectives for strategic human resource management. Journal of Management 18(2): 295–320.

[14] Tavares S, Caetano A (1999). A Emergência da Gestão de Recursos Humanos Estratégica. In Caetano A (Coord), Mudança Organizacional E Gestão de Recursos Humanos Lisboa: Observatório do Emprego e Formação Profissional.

[15] Prahalad CK, Hamel G (1990). The core competence of the corporation. Harvard Business Review 90(3): 79–91.

[16] Peretti J-M (1998). Recursos Humanos (2nd edn). Lisboa: Sílabo.

[17] Dussault G, Souza L (1999). Gestão de Recursos Humanos Em Saúde Canadá: Universidade de Montreal.

[18] Biscaia A, Conceição C, Martins J, Ferrinho P (2003). Política e gestão dos recursos humanos na saúde em Portugal – controvérsias. Revista Portuguesa de Clínica Geral 19: 281–289.

[19] Gregório J, Lapão L (2003). Uso de cenários estratégicos para planeamento de recursos humanos em saúde: O caso dos farmacêuticos comunitários em Portugal 2010–2020. Revista Portuguesa de Saúde Pública 30(2): 125–142.

[20] Dussault G, Fronteira I (2010). Recursos Humanos Para a Saúde (RHS): Plano Integrado No Plano Nacional de Saúde 2011–16 (PORTUGAL) Lisboa: Alto Comissariado da Saúde.

[21] Administração Central do Sistema de Saúde (2009). Estudo de Necessidades Previsionais de Recursos Humanos Em Saúde (Médicos) Lisboa: ACSS.

[22] Santana P coord (2013). Estudo de Evolução Prospectiva de Médicos No Sistema Nacional de Saúde [Relatório Final] Coimbra: Universidade de Coimbra.

[23] Observatório Português do Sistema de Saúde (OPSS) (2003). SAÚDE: Que Rupturas? Relatório de Primavera 2003 Lisboa: Escola Nacional de Saúde Pública.

[24] Decreto-Lei no. 19/88 (1988). Aprova a lei de gestão hospitalar. Diário da República 1. Série, no. 248–(20) 21–01.

[25] Decreto-Lei no. 177/09 (2009). Regulamenta a carreira médica. Diário da República 1. série no. 149 04–08.

[26] Decreto-Lei no. 117/98 (1998). Instituindo o regime remuneratório experimental. Diário da República, 1. Série-A no. 103 05–05.

[27] Rodrigues J, Biscaia L, Ministro L, Pisco L (1998). Monitorização da Qualidade Organizacional Dos Centros de Saúde Lisboa: Direcção Geral da Saúde.

Alamuri Suryanarayana

6 Critical role of managerial competencies in productivity enhancement interventions: an HRM perspective

Abstract: Of all the problems faced by management, *productivity and its enhancement* must surely be ranked as one of the most intractable. However, unfortunately, there is no single theory relating to productivity per se that can completely and unqualifiedly be accepted as accounting for all the known facts, and there is no definite set of prescriptions that are unequivocally supported by research data. Who is responsible for "productivity"? Is the individual worker expected to possess and provide his own competencies, or are increasing productivity levels a function of work groups? Or is it the responsibility of management and leadership? Answers to these questions are not simple. Some researchers and authors see productivity as being contained within the individual, while others view it as arising from sources outside the individual. In this chapter, we examine both points of view and show that all *interventions* aimed at productivity as well as its improvements encompass forces both within and external to the individual member of any organization. Developing and improving *competencies* to raise the productivity bar is required not just for individuals who plan to enter managerial positions or who currently manage organizations. It is needed to help anyone and everyone who wants to manage many aspects of life and relationships more and better. We need to change our behavior to acquire and develop competence, and to be savvier in our relationships with different kinds of people. People's productivity rises if they improve their social and emotional intelligence. Fostering the development of such intelligence and coming up with an appropriate set of individual, interpersonal, and group interventions for registering higher and ever-increasing levels of productivity in organizations from a human resource management perspective is the main theme of this chapter.

6.1 Introduction

Management is the principal activity that *makes a difference* in how well organizations serve people affected by them. How successfully an organization achieves its objectives and satisfies social responsibilities depends to a large extent on its managers. If managers do their jobs well, an organization will probably achieve its goals and become *productive.* How well managers do their jobs – managerial performance – is the subject of much debate, analysis, and confusion in the USA and many other countries. So is organizational performance – the measure of how well organizations do

https://doi.org/10.1515/9783112206775-006

their jobs. In this context, we need to discuss many different concepts and criteria for evaluating managers and organizations leading to their ultimate productivity.

Peter F. Drucker, one of the most respected writers on management, suggested two concepts: efficiency and effectiveness. As he puts it, *efficiency* means "doing things right" and *effectiveness* means "doing the right thing." A manager who is both efficient and effective is said to be *productive*. Efficiency – the ability to do the things right – is an input–output concept. An efficient manager is one who achieves outputs, or results, that measure up to the inputs (human resources, materials, and time) used to achieve them. Managers who are able to minimize the cost of the resources needed to achieve goals act efficiently. Effectiveness, in contrast, involves choosing the right goals. A manager who selects an inappropriate goal – say, producing mainly large cars when demand for small cars is soaring – is an ineffective manager, even if the large cars are produced with maximum efficiency. Managers at General Motors (GM) learned this lesson the hard way. When the demand for fuel-efficient, smaller cars increased in the 1970s, GM ignored the competition created by the Japanese and Germans, believing that the trends were an aberration and that Americans, loyal to American products, would not continue to buy foreign cars. As a consequence, they continued to produce large, fuel-inefficient cars and in so doing lost enormous competitive ground to these new rivals. *No amount of efficiency can make up for a lack of effectiveness.*

In fact, Drucker says, effectiveness is the key to an organization's productivity and success. Before we can focus on doing things efficiently, we need to be sure we have found the right things to do.

In this context, human resource management (HRM) is critical for all managers of an organization, for at least two reasons. First, it can improve *productivity*, which improves an organization's financial health. Second, it can help organizations meet customers' *competitive priorities*. Productivity, the ratio of output to input, is a measure of a manager's or an employee's efficiency in using an organization's scarce resources to produce goods and services. The higher the numerical value of this ratio, the greater the efficiency. Ernst & Young managers use "hoteling" to affect both parts of this ratio. They seek to cut inputs (space costs) and to boost the output of traveling accountants. To understand the connection between productivity and efficiency, we need to look at Table 6.1 and try to visualize it as a series of strategic control points; the points refer to junctures at which a major change occurs.

In filling a customer order, for example, strategic control points would occur when the purchase order becomes an invoice, when an inventory item becomes an item to be shipped, and when an item to be shipped becomes part of a truckload of goods to be delivered. Any of these strategic control points is a potential source of confusion and inefficiency, as work is passed from one set of workers to another.

An unclear form or a confusing policy – say, for handling out-of-stock items – creates the risk that orders will be lost or mishandled, wasting valuable time, money, or energy. From this vantage point, the operations system looks like a sieve that can leak

Table 6.1: Characteristics of products and services.

Dimension	Product	Service
Output	Tangible	Intangible
Output consumption	Can be stored over time	Immediate; cannot be stored
Nature of work	Producer intensive	Labor intensive
Customer contact	Minimal, indirect	Direct
Customer participation	Little or none	Essential

Source: Adapted from [1].

valuable resources unless it is managed efficiently. Productivity offers one measure of this efficiency. For example, assume that a legal clinic with eight lawyers (the input) produces output consisting of 100 client consultations per day. Productivity would equal 100/8 or 12.50. Assume that a second legal clinic next door has 15 lawyers handling 125 consultations per day. The productivity ratio would be 125/15 or 8.33. The smaller clinic has a higher productivity ratio on a quantitative basis, which may or may not reflect anything about the quality of its output, of course!

Types of productivity ratios: There are two basic types of productivity ratios. The first, *total productivity*, relates the value of all output to the value of all input, using the ratio of total output to total input. The second, *partial productivity*, relates the value of all output to the value of major categories of input, using the ratio of total output to partial input. The legal clinic is an example of a partial productivity ratio, called a *labor/employee productivity index* or *output per work-hour ratio*. Most productivity measures cited by economists and business executives are, in fact, labor/employee productivity indexes since human resources are one of the greatest ongoing costs for most organizations. Other partial productivity ratios measure the amount of scrap (wasted materials); the number of units that must be reworked or fixed before they meet quality standards; cycle time, the length of time to perform an operation; and downtime, the unproductive time spent retooling a production line or waiting for customers. Any of these measures gives an indication of whether resources are being used to good advantage or wasted.

Uses of productivity ratios: Productivity ratios can be calculated for a specific time period, which measures the efficiency of operations at that time, or they can be compared with other ratios over time, as a measure of gains or losses in productivity. For example, between 1988 and 1992, Mexican manufacturing employees recorded an approximate 6.5% average annual productivity increase, compared to 3% average annual gain over the same period by US-based manufacturing employees. In recent years, US manufacturers have tried to boost productivity by closing plants, downsizing, laying off production workers, and selling off falling or unwanted businesses.

Still, as an economic system, the USA lags behind Japan, South Korea, Great Britain, Norway, Sweden, France, and other countries in *productivity growth*. Many government officials and executives are searching for solutions to this problem.

Most companies develop competitive problems when managers lose sight of their organization's primary reason for being: *to produce quality products and services that consumers want at prices that seem reasonable.* This relates back to measures of organizational effectiveness: *the ability to set the "right" goals, ones that build on organizational strengths and meet the needs and wants of potential consumers.* Of course, needs and wants of an individual vary widely, as do price perceptions. Rather than try to be all things to all consumers, effective managers make strategic decisions about how their organizations can best meet their customers' competitive priorities through their human resources, and then adjust their operations accordingly.

After all, customers use an organization's products and services in their own competitive context, while organizations must keep in mind major competitive priorities such as pricing, quality level, quality reliability, and flexibility. Many experts say the problem is the emphasis on productivity itself. They charge that in trying to improve "the numbers" – quantitative measures of productivity – too many US managers have focused on capital investment in automation as a way to reduce labor costs. This short-term focus has caused them to overlook the benefits of investing in the organization's human capital – employees and their skills – and improving quality.

This emphasis is changing as managers at many organizations concentrate on finding the right mix of capital investment and human investment. One of the most important trends in HRM is the focus on increasing workforce literacy, knowledge and skills that relate directly to job performance. Another is the trend toward participative management and the use of self-managed work teams to improve productivity and quality simultaneously. Investment in human capital is increasingly important not only to manufacturing but also to the service and knowledge-oriented economy. As millions of people are displaced by technology, the downsizing of corporations, and competition from around the globe, issues of social responsibility of companies toward their employees become more crucial.

According to Peter Drucker, raising the productivity of service work is management's first social responsibility. He argues that we must work smarter. This involves making sure that continuous learning accompanies productivity gains and recognizing that knowledge workers and service workers learn most when they teach.

6.2 The critical role of managerial competencies

Predictions of the changes that will occur in the future are often notoriously wrong. No one doubts that the twenty-first century will continue to be characterized by chaotic, transformational, rapid-fire change. In fact, almost no sane person is willing to predict what the world will be like 50, 25, or even 15 years from now. Change is just too rapid and ubiquitous. The development of so-called *nanobombs* has caused some people to predict that personal computers and desktop monitors will land on the scrap heap of obsolescence within 20 years. The new computers will be a product of etchings on molecules lading to personalized data processors injected into the bloodstream, implanted in eyeglasses, or included in wristwatches. Tom Peters counseled managers that, due to the chaotic pace of change, "If you're not confused, you're not paying attention."

And the late Peter Drucker characterized the current environment thus: *people don't understand the world anymore, and the past is not sufficient to explain our current environment.* Almost everything is in flux, from our technology and methods of transacting business to the nature of education and the definition of the family. Despite all this change in our environment, there is something that has remained, and continues to remain, relatively constant. With minor variations and stylistic differences, what has not changed in several thousand years are the basic skills and competencies that lie at the heart of effective, satisfying, growth-producing, and productive human relationships. Freedom, dignity, trust, love, and honesty in relationships have always been among the goals of human beings, and the same principles that brought about those outcomes in the eleventh century are still at work in the twenty-first century. Despite our circumstances, in other words, and despite technological resources we have available to us, the same basic human skills still lie at the heart of effective human interaction. It is a fact that when everything is changing, change becomes unmanageable.

6.2.1 What are management competencies?

There are several defining characteristics of managerial competencies that differentiate them from other kinds of managerial characteristics and practices.

First, they are behavioral in nature. They are not personality attributes or stylistic tendencies. They consist of identifiable sets of actions that individuals perform that lead to certain productive outcomes. They can be observed by others, unlike attributes that are purely mental or are embedded in personality. Whereas people with different styles and personalities may apply the competencies differently, there are, nevertheless, a core set of observable attributes in effective skill performance that are common across a range of individual differences.

Second, they are controllable. The performance of these behaviors is under the control of the individual. Unlike organizational practices, such as "selectively hiring," or cognitive activities, such as "transcending fear," competencies can be consciously demonstrated, practiced, improved, or restrained by individuals themselves. They may certainly engage other people and require cognitive work, but they are behaviors that people can control themselves.

Third, they are developable. Performance and productivity can improve. Unlike IQ or certain personality or temperament attributes that remain relatively constant throughout life, individuals can improve their competency in skill performance to greater competency in management skills, and that outcome is the primary objective of any productivity enhancement interventions for an HR manager.

Fourth, they are interrelated and overlapping. It is difficult to demonstrate just one competency in isolation from others. They are simplistic, repetitive behaviors, but they are integrated sets of complex responses. Productive managers, in particular, must rely on combinations of skills to achieve desired results. For example, to effectively motivate others, competencies such as supportive communication, influence, empowerment, and self-awareness may be required. Productive managers, in other words, develop a constellation of competencies that overlap and support one another and that allow flexibility in managing diverse situations.

Fifth, they are sometimes contradictory or paradoxical. For example, core management competencies are neither all soft and humanistic in orientation nor all hard-driving and directive. They are oriented neither toward teamwork and interpersonal relations exclusively nor toward individualism and technical entrepreneurship exclusively. A variety of competencies are typical of the most effective managers, and some of them appear incompatible.

6.2.2 Improving management competencies

It is a bit unnerving that while average IQ scores have increased in the population over the last half-century, social and emotional intelligence scores have actually declined. In the general population, people are less skilled at managing themselves and managing others than they were 50 years ago [2]. Moreover, whereas the "technological float" has shrunk dramatically – that is, the time between the introduction of a new technology and its being copied and revised is constantly decreasing and is now measured in weeks rather than years – the "human float" has changed very little. It still takes about the same amount of time to develop behavioral skills and human competencies as it always has. No shortcuts or quick fixes have emerged, and the effort and practice that are required to become more emotionally intelligent and interpersonally skilled is substantial. Progress regarding how to cope with and manage issues relating to other people has not kept pace with technological progress, and it remains the biggest challenge all managers in general and more particularly HR managers face.

6.2.3 An approach to competency development for productivity enhancement

Successful management development, of course, is more than just following a recipe of sequential behaviors. Developing highly competent management competencies is much more complicated than developing skills such as those associated with a trade. They are: (1) linked to a more complex knowledge base than other types of competencies, and (2) inherently connected to interaction with other (frequently unpredictable) individuals. A standardized approach to welding or shooting free throws may be feasible, but no standardized approach to managing human beings is possible. The method that has been found to be most successful in helping individuals develop management competencies is based on social learning theory [3–5].

First, this approach marries rigorous conceptual knowledge with opportunities to practice and apply observable behaviors. It relies on cognitive work as well as behavioral work.

Second, individuals must be aware of their current level of skill competency and be motivated to improve upon that level in order to benefit from the model. Most people receive very little feedback about their current level of skill competency. Most organizations provide some kind of annual or semiannual assessment, but these assessments are almost always infrequent and narrow in scope, and they fail to evaluate performance in most critical competency areas.

Third, an application component is needed in the learning model. Most management skill training takes place in a classroom setting where feedback is immediate, and it is relatively safe to try out new behaviors and make mistakes. However, transferring learning to an actual job setting is often problematic. In summary, evidence suggests that a five-step learning model is most effective for helping individuals develop management competencies [6–9].

6.2.4 Competencies in personal, interpersonal, and group skills: an imperative for productivity enhancement

The message is simple and louder: from almost every perspective, competence in personal skills, interpersonal skills, and group skills is a critical prerequisite for productivity and success in management. Strong analytical and quantitative skills are important, but they are not sufficient. Productive managers must be able to work effectively with people. Research has shown that leadership and management skills fall into four clusters or categories. To be an effective manager, in other words, individuals must be competent in: (1) clan skills, or a focus on collaboration; (2) adhocracy skills, or a focus on creation; (3) market skills, or a focus on competition; and (4) "hierarchy" skills, or a focus on control. Clan skills include those required to build effective interpersonal relationships and develop others (e.g., building teamwork, communicating supportively). Adhocracy skills include those required to manage the future, innovate, and promote change (e.g., solving problems creatively, articulating an energizing vision). Market skills include those required to compete effectively and manager external relationships (e.g., motivating others, using power, and influence).

Hierarchy skills include those required to maintain control and stability (e.g., managing personal stress and time, solving problems rationally) [10]. Because circumstances are constantly changing and expectations for performance and productivity are continually escalating, the traditional definition of management has become outmoded and irrelevant. Effective managers and leaders do largely the same things in dealing effectively with constant change and constant stability. Twenty-first *century managers need to develop competencies that will enhance our ability to be both* leaders *and* managers.

6.3 Contents of this book chapter

This chapter focuses on the skills that research has identified as being critically important for successful management and leadership.

Section 6.1 deals with personal skills: developing self-awareness, managing personal stress, and solving problems analytically and creatively. These skills focus on issues that may not involve other people but instead relate to the management of the self, hence the term personal skills. The section touches upon clusters of related behaviors, not just one single, simple skill. These clusters of interrelated behaviors comprise the overall management skills under the section heading. It may be mentioned here that each cluster is related to and overlaps with other personal management skills, so each relies at least partially on the others to be applied productively and successfully.

Section 6.2 deals with interpersonal skills: building relationships by communicating supportively, gaining power and influence, motivating others, and managing conflict. These skills focus primarily on issues that arise in our interactions with other people. Overlap exists among these skills, of course, so that one must rely on parts of many skill areas to apply any one competence effectively.

Section 6.4 covers group skills: empowering and delegating, building effective teams and teamwork, and leading positive change. These competencies focus on key issues that arise when one is involved with groups of people either as a leader or as a member of the group. As with all the skills to be discussed in this chapter, overlap occurs among group skills as well as with personal and interpersonal skills.

Conversely, as we progress from personal to interpersonal to group skills, the core competencies developed in the previous skill area will help support successful application of the new skill area.

6.4 Improving personal competencies

6.4.1 Self-awareness

Most parts of Section 6.1 relate to skills in interpersonal or group interactions, but successful competency development in those areas will occur only if individuals have a firm foundation in self-awareness. In fact, there is an interesting paradox in human behavior: we can know others only by knowing ourselves, but we can know ourselves only by knowing others. Our knowledge of others, and therefore our ability to manage or interact successfully with them, comes from relating what we see in them to our own experience. If we are not self-aware, we have no basis for knowing how to recognize and understand others. Self-recognition leads to recognition and understanding of others. As Harris [11] puts it, "nothing is really personal that is not first interpersonal, beginning with the infant's shock of separation from the umbilical cord." What we know about ourselves comes only from the outside and is interpreted by the kind of experience we have had, and what we know about others comes only from analogy with our own network of feelings.

Companies have begun to discover the power of developing self-awareness among its managers. An awareness of how individuals differ in their emotional maturity, value priorities, and maturity of values, cognitive style, orientation toward change, and personality has helped many companies cope better with interpersonal conflicts, botched communications, breakdowns in trust, and misunderstandings. Not only does self-awareness training assist individuals in their ability to understand, and thereby manage, themselves, but it is also important in helping individuals develop understanding of the differences in others. Most people will regularly encounter individuals who possess different styles, different sets of values, and different perspectives than they do. Most workplaces are becoming more, not less, diverse. Self-awareness training can be a valuable tool in helping individuals develop empathy and understanding for the expanding diversity they will face in work settings. Self-awareness is a key component of and prerequisite for successful management.

Following are the *behavioral guidelines* relating to the improvement of self-awareness:

1. Identify your sensitive areas. Determine what information about yourself you are most likely to defend against.
2. Use the seven dimensions of national culture to identify differences between your own values orientation and that of individuals from other cultures, age categories, or ethnic groups.
3. Identify a comprehensive, consistent, and universal set of principles on which you will base your behavior. Identify the most important terminal and instrumental values that guide your decisions.
4. Expand your cognitive style, your tolerance of ambiguity, and your internal locus of control by increasing your exposure to new information and engaging in different kinds of activities than you are used to. Seek ways to expand and broaden yourself.
5. Enhance your emotional intelligence by consciously monitoring your own emotional responses and by practicing the reading of others' emotional cues.
6. Develop a healthy core self-assessment and positive self-regard by consciously capitalizing on your personal strengths and by highlighting and building on your accomplishments.

7. Engage in honest self-disclosure with someone who is close to you and accepting of you. Check out aspects of yourself that you are not sure of.
8. Keep a journal, and make time regularly to engage in self-analysis. Balance life's activities with some time for self-renewal.

6.4.2 Competency to manage personal stress

The best way to manage stress is to eliminate it through time management, work redesign, prioritizing, goal setting, and *small wins*. This strategy has permanent consequences, but it often takes an extended period of time to implement. Four kinds of stressors: time, encounter, situational, and anticipatory, cause negative physiological, psychological, and social reactions in individuals. These reactions are moderated by the *resiliency* that individuals have developed for coping with stress. Improving one's resiliency is an effective stress management strategy. When stressors are long lasting or are impossible to remove, coping requires the development of personal resiliency.

This is the capacity to withstand or manage the negative effects of stress, to bounce back from adversity, and to endure difficult situations [12]. Physiological resiliency is strengthened through increased cardiovascular conditioning and improved diet. Psychological resiliency and hardiness are improved by practicing a small-wins strategy and deep relaxation. Social resiliency is increased by fostering mentoring relationships and teamwork among coworkers. These strategies produce long-term benefits, but they also take quite a long time to implement. When circumstances make it impossible to apply longer-term strategies for reducing stress, short-term relaxation techniques can temporarily alleviate the symptoms of stress. These strategies have short-term consequences, but they can be applied immediately and repeated over and over again.

Eliminating sources of stress and developing resiliency to stress are the most desirable stress-management strategies as they have a permanent or long-term effect on our well-being. However, the occurrence of stressors is sometimes beyond our control, so it may be impossible to eliminate them. Moreover, developing resiliency takes time, so sometimes we must use temporary reactive mechanisms to maintain equilibrium. Although increased resilience can buffer the harmful effects of stress, we must sometimes take immediate action in the short term to cope with stress.

Implementing short-term strategies reduces stress temporarily so that longer-term stress-elimination or resiliency strategies are largely reactive and must be repeated whenever stressors are encountered because, unlike other strategies, their effects are only temporary. Five of the best-known and easiest-to-learn techniques are: (i) *muscle relaxation* and (ii) *deep breathing* which are physiological, and (iii) imagery and fantasy, (iv) rehearsal, and (v) reframing which are psychological. Muscle relaxation involves easing the tension in successive muscle groups. Deep breathing is done by taking several successive slow, deep breaths and holding them for five seconds.

Imagery and fantasy eliminate stress temporarily by changing the focus of one's thoughts. Imagery involves visualizing an event using "mind pictures." An increasingly common practice for athletes is to visualize successful performance or to imagine themselves achieving their goal. Research has confirmed both the stress-reduction advantages of this technique as well as the performance enhancement benefits (e.g. [13]). Using a *rehearsal* technique, people work themselves through potentially stressful situations, trying out different scenarios and alternative reactions.

Appropriate reactions are rehearsed, either in a safe environment before stress occurs or "off-line," in private, in the midst of a stressful situation. Reframing involves temporarily reducing stress by optimistically redefining a situation as manageable. It serves as a key to developing "hardiness" and "emotional intelligence."

6.4.3 Competency to solve problems analytically and creatively

In the twenty-first century, almost no manager or organization can afford to stand still, to rely on past practices, and to avoid innovation. In a fast-paced environment in which the half-life of knowledge is about 3 years and the half-life of almost any technology is counted in weeks and months instead of years, creative problem solving is increasingly a prerequisite for success. The digital revolution makes the rapid production of new ideas almost mandatory. This is not to negate the importance of analytical problem solving, of course.

The quality revolution of the 1980s and 1990s taught us important lessons about carefully prescribed, sequential, and analytical problem-solving processes. Error rates, response times, and missed deadlines dropped dramatically when analytical problem solving was institutionalized in manufacturing and service companies. Problem solving is a competency that is required of every person in almost every aspect of life. Seldom does an hour go by without an individual's being faced with the need to solve some kind of problem. The manager's job, in particular, is inherently a problem-solving job.

If there were no problems in organizations, there would be no need for managers. Therefore, it is hard to conceive of an incompetent problem solver succeeding and being productive as a manager. Effective managers are able to solve problems both analytically and creatively, even though different competencies are required for each type of problem. Managers use analytical problem solving many times each day. Creative problem solving occurs less frequently but separates career successes from career failures, heroes from goats, and achievers from derailed executives. It can also have a dramatic impact on organizational effectiveness. A great deal of research has highlighted the positive relationship between creative problem solving and successful organizations [14]. Managers need to foster creative problem solving and innovation among their coworkers.

6.5 Improving interpersonal personal competencies

6.5.1 Competency to build positive interpersonal relationships

A great deal of research supports the idea that positive interpersonal relationships are key to creating positive energy in people's lives mycitebib:c06-TS73a,bib:c06-TS73b. When people experience positive – even if they are just temporary encounters – they are elevated, revitalized, and enlivened. Positive relationships create positive energy. The effects of positive relationships are much stronger and longer lasting than just making people feel happy or uplifted, however. Several benefits occur because positive relationships actually strengthen the immune system, the cardiovascular system, and the hormonal system [15–17]. People's intellectual capacities are actually broadened (mental acuity expands), they learn more and more efficiently, and they make fewer mental errors when experiencing positive relationships [18]. Creativity and innovation, as well as the system's capacity to adapt to change, are substantially higher when positive relationships characterize the workforce [19].

On the other hand, in study after study, communication problems are identified as the single biggest impediment to positive relationships and positive performance in organizations [20, 21]. The ability to communicate supportively is the single most important skill that effective managers must possess. The most important barriers to effective communication in organizations are interpersonal. Much technological progress has been made in the last two decades in improving the accuracy of message delivery in organizations, but communication problems persist among people, regardless of their relationships or roles. A major reason for these problems is that a great deal of communication does not support a positive interpersonal relationship. Instead, it frequently engenders distrust, hostility, defensiveness, and feelings of incompetence and low self-esteem.

6.5.2 Building a strong power base and using influence wisely as a managerial competence

It should come as no surprise that many authorities argue that the effective use of power is the most critical element of management. One such authority, Warren Bennis, seeking the quintessential ingredients of effective leaders, found that most influential leaders in all walks of our society shared one significant characteristic: they made others feel powerful. They became powerful as they learned how to build a strong power base in their organizations or institutions.

They were influential because they used their power to help peers and subordinates accomplish exceptional tasks. It requires no particular power, skill, or genius to accomplish the ordinary. But it is difficult to do the truly unusual without political clout [22]. Two competencies, gaining power and translating that power into influence, must be developed if one is to maximize one's potential as a power holder. A strong person in a weak position and a weak person in a strong position are both at a disadvantage. Ideally, one should grow into a strong person in a strong position to become and remain productive.

6.5.3 The skill of interpersonal conflict management as a management competency

Conflict is a difficult and controversial topic because in most cultures it has negative connotations as it runs counter to the notion that we should get along with people by being kind and friendly. Although many people intellectually understand the value of conflict, they feel uncomfortable when confronted by it. Their discomfort may result from a lack of understanding of the conflict process as well as from a lack of training in how to handle interpersonal confrontations effectively. One of the leading causes of business failure among major corporations is too much agreement among top management. A conflict over issues is not only likely within top-management teams but also valuable. Such conflict provides executives with a more inclusive range of information, a deeper understanding of the issues, and a richer set of possible solutions.

Research evidence suggests that the alternative to conflict is usually not disagreement, but apathy and disengagement. In fast-paced markets, successful strategic decisions are most likely to be made by teams that promote active and broad conflict over issues without sacrificing speed. The key to doing so is to mitigate interpersonal conflict [23]. Managers need to focus more on the effective implementation of a specific conflict management approach that is both the most effective, all-purpose tool and the most difficult to use comfortably and skillfully – collaborative problem solving. It takes little skill to impose one's authority or positional power that can be exercised over another person either: (i) to withdraw from a confrontation, or (ii) to split the difference between opponents, or (iii) to abandon one's position at the slightest sign of opposition. It calls for equipping oneself with the necessary behavioral constellations for resolving an interpersonal confrontation involving complaints and criticisms by using a problem-solving approach and creating win–win situations.

6.6 Improving group competencies

6.6.1 Empowering and delegating as a management competency

Empowerment is based on a set of assumptions that contrast with those normally made by managers. It means helping to develop in others a sense of self-efficacy, self-determinism, personal control, meaning, and trust. The present business environment is not particularly compatible with the principles of managerial empowerment. Because of the turbulent, complex, competitive circumstances that many organizations face, managers frequently experience a tendency to be less, rather than more, empowering. When managers feel threatened, they become rigid and seek more control over other people, not less. However, without empowered employees, organizations cannot succeed in the long run. Learning how to be a competent, empowering manager is therefore a critical skill for individuals who probably have a predilection not to practice empowerment. Empowerment means providing freedom for people to successfully do what they want to do, rather than getting them to do what you want them to do. Managers who empower people remove controls, constraints, and boundaries for them instead of motivating, directing, or stimulating their behavior. Empowering others, however, can lead to dilemmas.

On the one hand, evidence shows that empowered employees are more productive, more satisfied, and more innovative, and they create higher-quality products and services than unempowered employees [24–28]. Organizations are more effective when they have an empowered workforce [29–31]. Producing a sense of empowerment in others and delegating in a way that empowers subordinates also brings desirable outcomes for organizations and employees. Empowered employees are more proactive and innovative, persistent in their work, trustworthy, interpersonally effective, and intrinsically motivated and have higher morale and commitment than employees who are not empowered.

6.6.2 Building effective teams and team work in work teams as a management competency

All of us are members of multiple teams – at work, at home, and in the community. Teams are becoming increasingly prevalent in the workplace and in the classroom because they have been shown to be powerful tools in improving the performance and productivity of individuals and organizations.

Consequently, it is important to become proficient in leading and participating in teams. It is obvious that merely bringing people together and giving them an assigned task does not make them a team. There are three types of team skills: diagnosing and facilitating team development, leading a team, and being an effective team member. There exists a strong relationship among these three key competencies with highly productive and effective team performance. But to be skillful, managers need to hone their ability to perform each of these skills competently. Developing team skills is important because of the tremendous explosion in the use of teams in work organizations over the last decade or so. Empowered teams, autonomous work groups, semiautonomous teams, self-managing teams, self-determining teams, crews, platoons, cross-functional teams, top-management teams, quality circles, project teams, task forces, virtual teams, emergency response teams, and committees are all examples of the various manifestations of teams and teamwork that appear in the scholarly literature, and research has been conducted on each of these team forms.

However, managers need to hone and develop skills that are relevant in most or all of these kinds of situations, whether as team builders, team leaders, or team members. One management consultant, Tom Peters [10, p. 306], even asserted:

> Are there any limits to the use of teams? Can we find places or circumstances where a team structure doesn't make sense? Answer: No, as far as I can determine. That's unequivocal, and meant to be. Some situations may seem to lend themselves more to team-based management than others. Nonetheless, I observe that the power of the team is so great that it is often wise to violate apparent common sense and force a team structure on almost anything.

One reason for the escalation in the desirability of teamwork is that increasing amounts of data show improvements in productivity, quality, and morale when teams are utilized. Many companies have attributed their performance improvements directly to the institution of teams in the workplace [32–36].

6.7 Concluding comments and summary

A person's productivity rises if he improves his social and emotional intelligence. John Holt [8, p. 165] succinctly summarizes this by equating the skill to enhance one's productivity to intelligence.

> When we talk about intelligence, we do not mean the ability to get a good score on a certain kind of test or even the ability to do well in school; these are at best only indicators of something larger, deeper, and far more important. By intelligence, we mean a style of life, a way of behaving in various situations. The true test of intelligence is not how much we know how to do, but how we behave when we don't know what to do.

Fostering the development of such intelligence and coming up with an appropriate set of individual, interpersonal, and group interventions for registering higher and higher levels of productivity in organizations from a human resource perspective is the main theme of this chapter.

References

[1] Adam EEJ, Elbert RJ (1989). Production and Operations Management (4th edn.). Englewood Cliffs, N.J: Prentice Hall, 7.

[2] Goleman D (1998). Working with Emotional Intelligence New York: Bantam.

[3] Bandura A (1977). A Social Learning Theory Englewood Cliffs, NJ: Prentice Hall.

[4] Boyatzis RE (1996). Consequences and Rejuvenation of Competency Based Human Resource and Organization Development.

[5] Davis TRV, Luthans F (1980). A social learning approach to organizational behavior. The Academy of Management Review 5(2): 281–290.

[6] Cameron KS, Whetten DA (1984). A model for teaching management skills. Organizational Behavior Teaching Journal 8: 21–27.

[7] Kolb DA (1984). Experiential Learning: Experience as the Source of Learning and Development Englewood Cliffs, NJ: Prentice Hall.

[8] Vance CM (1993). Mastering Management Education Newbury Park, CA: Sage.

[9] Whetten DA, Cameron KS (1983). Management skill training; A needed addition to the management curriculum. Organizational Behavior Teaching Journal 8: 10–15.

[10] Cameron K, Quinn RE (2006). Diagnosing and Changing Organizational Culture: Based on the Competing Values Framework Beijing: China Renmin University Press.

[11] Harris M (1981). America Now: The Anthropology of a Changing Culture New York: Simon & Schuster.

[12] Masten AS, Reed MJ (2002). Resilience in Development. In: Handbook of Positive Psychology, Snyder CR, Lopez SJ (eds) New York: Oxford University Press, 74–88.

[13] Deepak MD (1995). Creating Health: How to Wake up the Body's Intelligence? New York: Houghton Mifflin.

[14] Sternberg RJ (ed) (1999). Handbook of Creativity Cambridge: Cambridge University Press.

[15] Dutton JE (2003). Energize Your Workplace: How to Build and Sustain High-quality Relationships at Work San Francisco, CA: Jossey-Bass.

[16] Dutton JE, Heaphy ED (2006). The Power of High-quality Connections. In: Positive Organizational Scholarship: Foundations of a New Discipline, Cameron KS, Dutton JE, Quinn RE (eds) San Francisco: Berrett-Koehler, 263–278.

[17] Reis HT, Gable SL (2003). Toward a Positive Psychology of Relationships. In: Flourishing: The Positive Person and the Good Life, Keyes CL, Haidt J (eds) Washington, DC: American Psychological Association, 129–159.

[18] Fredrickson BL (2001). The role of positive emotions in positive psychology: The broaden-and-build theory of positive emotions. American Psychologist 56: 218–226.

[19] Gittell J, Cameron K, Lim S (2006). Relationships, layoffs, and organizational resilience. Journal of Applied Behavioral Science 42(3): 300–329.

[20] Carrell LJ, Willmington SC (1996). A comparison of self report and performance data in assessing speaking and listening competence. Communication Reports 9: 185–191.

[21] Thorton BB (1966). Resolving international conflict: Culture and business strategy. Business Strategy Review 7: 51–68.

[22] Bennis W (2003). On Becoming a Leader Cambridge, MA: Perseus Books Group.

[23] Eisenhardt KE, Kahwajy JL, Bourgeois LJ III (1997). Conflict and strategic choice: How top management teams disagree, California Management Review 39(2) (Winter 1997) : 42–62. Reprinted in: Hambrick, Nadler, Tushman (eds), Navigating Change: How CEOs, Top Teams, and Boards Steer Transformation, Harvard Business School Press.

[24] Greenberger DB, Strasser S (1991). The role of situational and dispositional factors in the enhancement of personal control in organizations. Research in Organizational Behavior 13: 111–145.

[25] Kanter RM (1983). The Change Masters: Innovations for Productivity in the American Corporation New York: Simon and Schuster.

[26] Sashkin M (1982). A Manager's Guide to Participative Management New York: American Management Association.

[27] Sashkin M (1984). Participative management is an ethical imperative. Organizational Dynamics 12: 4–22.

[28] Spreitzer GM (1992). When Organizations Dare: The Dynamics of Individual Empowerment in the Workplace Ph.D. dissertation, University of Michigan.

[29] Conger JA, Kanungo RN (1988). The empowerment process. Academy of Management Executive 13: 471–482.
[30] Gecas V (1989). The social psychology of self-efficacy. Annual Review of Sociology 15: 291–316.
[31] Thomas KW, Velthouse BA (1990). Cognitive elements of empowerment: An interpretative model of intrinsic task motivation. Academy of Management Review 15: 666–681.
[32] Cohen SG, Bailey DE (1997). What makes teams work: Group effectiveness research from the shop floor to the executive suite. Journal of Management 23(3): 239–290.
[33] Guzzo RA, Dickson MW (1996). Teams in organizations: Recent research on performance and effectiveness. Annual Review of Psychology 47: 307–338.
[34] Hamilton BH, Nickerson JA, Owan H (2003). Team incentives and worker heterogeneity: An empirical analysis of the impact of teams on productivity and participation. The Journal of Political Economy 111(3): 465–497.
[35] Katzenbach JR, Smith DK (1993). The Wisdom of Teams: Creating the High-performance Organization New York, Harper Business.
[36] Senge P (1991). The Fifth Discipline New York: Doubleday.

Sara Maria Mendes FERNANDES and Carolina Feliciana MACHADO

7 Competencies management in an SME: a lever for greater productivity

Abstract: In such a competitive market, in which we are integrated, it is important to create sustainable bases for companies to achieve their goals. Therefore, it is necessary to monitor employee performance and develop their skills to provide the company with talent suited to the roles in order to achieve the levels of success desired by the organization. This chapter presents methods, techniques, and instruments used for competency planning in a private small or medium-sized enterprise.

The main objective of this chapter is to verify to what extent the planning and consequent preparation of competency management contribute to the good functioning of a small or medium-sized company. As well as verifying the influence that this planning will focus on acquiring and developing skills for each employee, to gradually increase their satisfaction and motivation in performing their role. Therefore, the better the planning of skills, the greater the chance for the company to succeed, and the better the performance and use of each person's capabilities. The company's production and its resulting success depend on the development of these same skills.

A quantitative approach research was carried out. A scale questionnaire was used as the research instrument. The research has shown that skills management in this company is well implemented, which translates into satisfaction for the majority of workers! In addition, the employees presented a development of skills derived from training actions and lectures according to the needs of each role. What is essential for a good level of work and cooperation to improve performance in order to make the company grow in the market. The main reasons for this success come from the way the leader deals with the problems presented by his employees and the way he helps them to become increasingly competent and capable.

7.1 Introduction

Competency management has been highlighted as a differential management model, which makes a difference in small or medium-sized enterprises (SMEs) when it comes to their competitive strength in the market. The acquisition of certain skills is a determining factor both in employee performance and company success.

According to Brandão and Bahry [1] and Mlaouhi et al. [2], competency management is a model that uses the concept of competency to guide various organizational models, especially those focused on people management. Using different strategies and instruments that aim to identify, develop, and mobilize skills necessary for the achievement of organizational objectives.

https://doi.org/10.1515/9783112206775-007

Thus, the main objective of this chapter is related to the analysis of the importance of planning in the management of competencies in SMEs.

The objectives of this chapter will involve analyzing the company's evolution with the help of competency management and its influence on the company's progression or regression. Know the utmost importance that competency management has in a company, whether it can be discarded or not, whether it can be postponed or not. Realize the importance of managing an employee's skills in the workplace. Find out what advantages and consequences planning brings to organizations' competency management, more specifically in *Smile Company,*[1] an SME.

In the first section of the work, a literature review is presented, where competencies management in companies is developed, establishing an approximation with the concept of competencies management and developing what is understood as the effectiveness of the process. The theme of small and medium-sized companies is introduced in a second point, highlighting its main characteristics. To conclude this part, the objective is to address the importance of skills management in SMEs to develop the present study. Next comes Section 7.2, which consists of the methodology where the company is presented, the resources used, the choice of the type of study, and the discussion and presentation of the results.

7.2 The importance of competency management in SMEs

7.2.1 Competency management in companies

7.2.1.1 Approach to the concept of competency management

Competency management applied in companies serves mainly to retain employees, and is seen as a very attractive and fair form of management [3].

Living in the world we live in, which is constantly changing, and taking into account the current business environment, it is important to understand that attitudes of involvement and behaviors are more important than technical knowledge itself. Nowadays, companies give less interest to employees who know the theory of a certain subject compared to those who demonstrate learning capacity and experience. Thus, there are transversal skills and specific skills [4]. Ceitil [5] stated that the concept of transversal skills is distinguished from specific skills (associated with a specific function, profession, or job) due to their transversality (non-contextualization) and transferability (acquired in one context and capability of being exercised in different

1 For reasons of anonymity the real name of the company is replaced by 'Smile Company'.

contexts). Knowledge is the key word to obtain a competitive advantage in companies. The success of organizations depends largely on their concern for personal and organizational knowledge.

As McClelland [6] concluded, technical skills and knowledge are not indicators of either professional performance or success in life, and the best indicator is a person's ability to react in an unexpected moment, as it demonstrates the ability to be reactive and effective in the face of adverse situations.

Competency management is nothing more than an organizational learning process, planned based on operational management and definition of strategies that allow the company to obtain better results, through its greatest strength, which is its people.

In this same vein, Brito [7] speaks of the concept of competencies as "the ability to manage the set of knowledge (formal knowledge), skills (know-how), and attitudes (wanting to do) of the worker to fulfill the mission, business and strategy of the organization and project the essential competency into the future." Competencies consist of personal capabilities that enable people to perform their duties successfully, achieving results or completing tasks efficiently and effectively [8].

7.2.1.2 Effectiveness of competency management

Competency management is made up of some processes, such as performance evaluation, where attention must be paid to which competencies exist and which are necessary to develop for organizational success [9]. There is also a skills development plan that allows people to develop their skills, being critical and seeking to provide feedback, to give wings to the possibility of continuous development on the part of each employee. And finally, skills management, and the better the analysis and description of functions are structured and constructed, the better this skills management can be carried out [10]. If a manager knows how to manage skills well, it will be easier to take precautions and control conflicts, strengthening team spirit or trying to develop each individual in the company. If the manager is unable to do all of this, he or she can and should delegate. The skills management referred to by Jardim [11] implies appealing to the promotion of practical intelligence, based on the mobilization of theoretical, practical, and methodological knowledge, "the theoretical knowledge that defines reality from within itself, when invested in action, unfolds into technical knowledge (defines what should be done) and methodological knowledge (how it should be done)."

In view of the above, competency management is also present in recruitment and selection and is absolutely linked to training management and rewards management [10].

Organizations' concern about hiring candidates with certain specific skills to efficiently perform a given role is not new. Taylor already warned, at the beginning of the last century, about the need for companies to hire people according to their capabilities, highlighting that the demand for candidates with skills or talents exceeded the supply. Our behaviors result from a bipolar form of the combination of two dimensions: the dimension of technical and functional skills; knowing how to be and knowing how to behave. This is why "I exist therefore I feel" is so important, in other words, emotional intelligence.

According to Cassol et al. [12], an employee with social skills has the ability to adapt their behaviors to different everyday situations. With the constant change in the market and competitiveness of companies, it becomes imperative that employees are able to develop competitive capabilities, such as social skills that will differentiate them. These types of skills are associated with the organizational strategy that involves the development of the individual and management.

And for this reason, managers should focus more on the personal and technical skills of their employees, always trying to improve them.

7.2.2 Small and medium-sized enterprises

7.2.2.1 The concept of SMEs

SMEs are companies that employ less than 250 employees [13, 14]. They can be micro (which are those with the most significant weight), small, or medium-sized companies and are the dominant base of each country in terms of the business area. In Portugal, this definition is regulated by Decree-Law 372/2007, of November 6.

SMEs create new jobs, employing around one third of the European workforce. Since they are in greater numbers, they are also the ones who create the most jobs. The ease with which SMEs can change makes changes much easier [14]. Large companies take much longer to change and alter because it is much more complex.

An SME, as it is small, can easily restructure itself immediately, so the response to customers is much faster as it is easier for them to adapt to the market that is constantly changing. They add greater flexibility to the industrial structure, promoting great dynamism in the economy [14, 15].

7.2.2.2 Main characteristics of SMEs

The main characteristics of these companies are that they have simpler organizational structures, where an employee can do everything, with great flexibility and adaptation to different job positions. There is no specific individual for just one role, there is more versatility. This aspect helps them to easily adjust to change to meet all the needs presented by their customers. Many are family businesses, which, according to

Roffia and Moracchiato [16] and Mandl [17], implies the existence of family involvement in company decisions.

Another characteristic that is the opposite of large companies is that SMEs usually have limited resources, whether financial or human [14, 15]. In the case of a shortage of human resources, this happens because people tend to want to join large companies, so the tendency to apply to small companies is small. The truth is that the idea that large companies can offer better opportunities for career development and training and that SMEs cannot is still very present, which is not always the reality and in many cases is the complete opposite. For example, employees of SMEs are able to acquire new roles and experience notable growth within the company much faster than employees of large companies. The opportunity to try out new roles, new positions and even learn more is in small and medium-sized companies where there is almost no power distance, the structure allows informality in work relations and management practices. For this reason, an employee can easily speak to the company's CEO, which does not happen in large companies [18].

They are also companies with great flexibility in adapting to the job market, they accept change and innovation more easily because they have the need to constantly adapt. For this reason they are constantly doing market research to acquire more information about their target audience. What helps SMEs a lot is the investment in technologies to help in this constant search to please the consumer and satisfy market needs [19]. They are strongly affected by internal and external factors, that is, as they do not have the strong stability of a large company, external factors such as labor laws and taxes affect SMEs much more.

7.2.3 The importance of competencies management in SMEs

Managing human resources always involves managing people, which according to Chiavenato [20] for a company to be successful requires considering its entire organizational structure, always bearing in mind that human capital is fundamental to this. Managing people involves managing conflict. Knowing that conflicts are not nonexistent, an HR manager has to be able to manage them because we live in an information society, so we have to try to transform information into knowledge [21] and finally know how to use communication to be able to manage skills.

If we have a completely disintegrated person, who doesn't know the organization, who doesn't feel part of the puzzle, or who does not feel part of the team, the tendency is to have lower performance, as the person feels dissatisfied, or completely unmotivated. If a person has a merely instrumental impairment, it may often have to do with the fact that they have not been well integrated. Considering that, with integration comes the cultural part where the objective will be to have the person feeling the organization as their own, and in an SME this dissatisfaction, of a single person, can influence an entire team because it is a much smaller environment, then it is necessary to stimulate everyone's satisfaction so that the company can truly reach its peak. That is why all human resources stages or activities presuppose evaluation. Quinn et al. [22] see the success of an organization as dependent on the delegation and division of tasks, so that all people can efficiently achieve their goals and objectives stipulated in the individual plan defined by HR managers, emphasizing that only in this way will they be able to achieve a common objective and achieve organizational goals.

Planning the competency management process consists of building a guide or plan that represents all the competencies required for each function in the company. Only then will it be possible to know what kind of skills employees have to acquire or demonstrate to successfully perform a respective function. This plan becomes a task map, which all companies should have, as this map is a guide to understanding which people to hire or what training to give to existing workers in the company, so that they are able to perform their role in the best possible way. SMEs need to be rigorous and demanding in all processes to be able to update themselves and meet market demands. To achieve this, employees must have and develop the skills required for their roles, ultimately streamlining the decisions they make on a daily basis, making their work more flexible in order to maintain the company's survival in the current market [23].

It is also with this task map that it can be verified, in performance evaluations, for example, that someone has skills that are not being well used and would fit better in another type of role. This would enable adjusting each person's profile to a more suitable role, managing careers. We, human beings, are not always able to see that what we do is not our vocation, and the fact that someone wants to change our position creates conflicts. However, after the acceptance phase, we were even able to agree that the role we were performing was not the most appropriate for our profile.

One of the important points for competency management is remuneration [24]. Many small businesses have already realized that one of the most efficient ways to recognize their employees is to reward them with monetary values in addition to their fixed salary, using, for example, goal-based management. This makes people self-motivated to do more for the company and also feel good about not accepting other job proposals. The competitive strength of SMEs is human capital and to achieve this, it is necessary to pay close attention to the profile of each person, the role they perform (checking whether or not it is appropriate), feedback and even rewards. Drawing up a fair and effective compensation plan is a fundamental step towards the success of this process.

Well-structured and appropriate skills management planning results in employee satisfaction, talent retention, and a positive impact on the company's image. It is also important that there is a plan that contains the necessary training to develop and prepare an employee to occupy a leadership position in the future of the organization. But care must be taken in its implementation, so that it achieves the desired goals. Therefore, one must be aware of the major stages to be covered from planning to execution and subsequent monitoring. Competency, ethics and responsibility are required at all stages. Skills can, in fact, be taught. This is where competency management planning comes in to define what type of training each employee really needs. This planning can be a follow-up to a performance evaluation where evidence emerges for a survey of training needs, integration or even the acquisition of new functions within the company.

Lana and Ferreira [25] highlight competency management as something that contributes significantly to a change in the strategic management vision of any company.

In view of the above, it would not be at all unreasonable to present the following hypothesis: "Effective management planning positively influences more efficient skills management." This is because this chapter will focus on planning competency management in SMEs as the basis of everything, as it helps the organization to prevent future events in order to find solutions to prevent them effectively. Not forgetting that, as human capital is smaller, it is necessary to retain and develop talent within the company to subsequently achieve success.

7.3 Competency management in an SME: empirical study

7.3.1 Theoretical-methodological assumptions

To develop the empirical part of this chapter, a questionnaire was applied to employees of the "Smile Company." In addition to general data about the company, employees were asked about how they feel about the use and benefit of their skills in their day-to-day work.

The questionnaire that was created was based on the work developed by Pereira [26], on "Competency management – development and institutionalization of a portfolio of key competencies in a higher education student organization." It was from this questionnaire that an adaptation was made for a new survey in Google Forms with some parts of it, which was later delivered to each employee.

It is worth adding that the company's director offered to help in this second stage of the work, as well as allowing full access to the company's facilities. His testimony and that of the other collaborators were the basic foundations that made the facts observed in this work possible. For data processing, SPSS was used to analyze the data with greater precision.

7.3.2 Smile Company

Smile Company is a textile company specialized in the production of knitted clothing for men, women, and children. Founded in 1994, the company is located in the city of Guimarães. Currently with around 80 employees, divided into different sectors of the company, "Smile Company" has a young, dynamic team, combined with over 20 years of experience in the textile sector.

Smile Company's mission is to produce and distribute textile products that aim to provide consumers with unique experiences, combining product quality with constantly updating design and machinery, always seeking to understand, satisfy, and exceed market expectations.

The company's vision is to consolidate a strong position in the national and international markets as a manufacturer of textile products, through a strong investment in the areas of design and machinery. With innovation always present, *Smile Company* shows the necessary capabilities and versatility to be a leader in the textile sector, developing partnerships and introducing innovative solutions.

Sustainability at "Smile Company" is practiced considering the environment, thus contributing to the interests of future generations. *Smile Company* uses the necessary resources responsibly, thus reducing the impact on the environment. It is a very successful and renowned company in the Guimarães area, both due to the humanity of the company's director towards its employees and in terms of market position. During the pandemic, this company also began to produce a large number of protective masks. The company had to adapt to the situation at the time, having purchased its machines for this purpose.

According to literature in the area of HRM in SMEs, there are two opposing perspectives: "small is beautiful" and "bleak house." *Smile Company* positions itself in the "small is beautiful" scenario where there is good people management, a "family" atmosphere, and close working relationships. It fits into a professional context that promotes harmonious working relationships and a spirit of unity through moments of conviviality such as Saint Martin celebrations, June festivals, and Christmas parties, where a family and intimate atmosphere is emphasized. It values human management and communication between the team. Although the CEO has not established an HRM department in the company (he himself is the human resources manager), he does not establish limits between himself and his employees because he believes that everyone works for the same purpose and no one is better than anyone else. Most of the time, those who come to the company from outside do not realize that he is the boss because he is among the machines and workers wearing the same clothes as them. The image of a boss does not exist at *Smile Company*, it is clear that the power distance does not exist. It provides flexibility to employees with paid absences without express justification, believing that people's word is enough.

7.3.2.1 Sociodemographic characterization

The participants in this study are part of the staff of the *Smile Company* (about 50% of workers). The inclusion criteria used were: working as an employee in the company's departments, excluding manufacturing and those who only interact with the company as intermediaries. In total, the study focused on 39 employees (21 women and 18 men), coming from 8 departments (managing partners, commercial technicians, packaging members, designers, controllers, sample making, accounting, and cutting members).

In order to assess whether there is any type of relationship between the gender of the respondents and the role performed, these two variables were crossed. This analysis resulted in the conclusion that, although the presence of women in management positions is more evident, there is a clear equivalence of men and women working in each department. However, it is worth highlighting that, despite this reality, the final decision always depends on what the CEO decides. This one always has the final word.

Regarding age, this variable was organized into five different groups, with a higher incidence in the 24 to 35 age group (35.9%) and in the 36 to 47 age group (28.2%) (Table 7.1).

Table 7.1: Distribution of employees by age.

Age	Frequency	Percentage
<24	6	15.4
24–35	14	35.9
36–47	11	28.2
48–59	6	15.4
>59	2	5.1
Total	39	100

It is important to highlight that despite the diversity of ages, it was found, during the study, that employees help each other and interact with each other without levels of superiority due to age differences, which will allow a transition of skills from one to the other, as well as influencing job satisfaction that will lead to the company's success.

Regarding the level of qualifications of respondents, there is a tendency for a bachelor's degree to be more common among managing partners, accounting staff, designers, and sales technicians (Figure 7.1). In other departments, it can be seen that there are different levels of qualifications in the same section, so it can be deduced that in the recruitment and selection process there is recognition of the experience that the employee has, and not just the technical skills that he has acquired. It can be seen that although there is not always the possibility of training, technical skills do not define the employee, but rather their know-how.

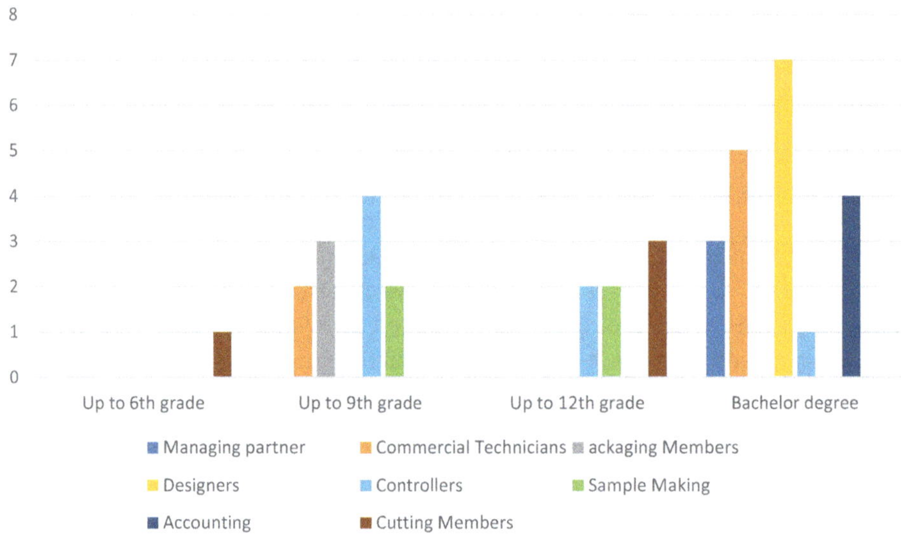

Figure 7.1: Qualification level by department (own elaboration).

7.3.3 Methodological instrument

Regarding data collection procedures, this study was based on the application of a satisfaction questionnaire survey regarding the use of each person's individual skills in their role. As stated before, the study focuses on a convenience sample corresponding to 39 members of the *Smile Company*. The questionnaires, duly adapted to the reality of the organization, were sent to each of the study participants' respective e-mails.

In this study, the CEO revealed how he evaluates the skills of his employees at various stages, from the moment he decides to hire the person to ongoing monitoring within the company.

7.3.4 Analysis and discussion of results

At this point, the objective is to establish a relationship between the information collected in the literature review and the results obtained in the empirical study.

It was previously mentioned that our behaviors result from the combination of two dimensions: the dimension of technical and functional skills; knowing how to be and knowing how to behave. This is why "I exist therefore I feel" is so important, in other words, emotional intelligence. That said, it is assumed that a manager must emphasize the personal and technical skills of his employees in order to always try to develop more capabilities.

According to Cassol et al. [12], an employee with social skills has the ability to adapt their behavior to different everyday situations. With the constant change in the market and competitiveness of companies, it becomes imperative that employees are able to develop competitive capabilities, such as social skills that will differentiate them. These types of skills are associated with the organizational strategy that involves the development of the individual and management.

In view of the above, it was decided that the questionnaire would focus on the personal skills of each employee precisely because there is a great need to plan these personal skills that are crucial to the success of SMEs.

This study began by addressing the methodological competencies (*selects and plans their activities and work methods, searches and selects information to address problems*, and *uses ICT in the exercise of their activities*). It was noticeable that not all employees plan and select their activities, or research and select information, nor do they all work with ICT, and this has to do with the fact that the exercise of their function requires or does not require the use of the same, but those who do and need to do so are successful in the action. This is in line with Section 7.2, where mention was made of the need for companies to equip and develop their employees with the skills necessary for their roles, ultimately streamlining the decisions they make on a daily basis and making their work more flexible to maintain the company's survival in the current market [23].

It is essential to understand that only those who perform a role with this type of requirement identify with this skill, even though this skill requires training and monitoring for the simple fact that the sources to be used for research require different approaches, treatments, and systematization so that knowledge is transformed into information of real value and, therefore, transactable to other members of other departments and even of the respective department so that the work flows naturally.

In a second point we focus a little on the issue of leadership (*ability to guide and motivate others to improve themselves* and *effectively delegate work tasks*), which was mentioned in Section 7.2.3, where the importance of having a plan that contains the necessary training to develop and prepare an employee to occupy a leadership position in the future of the organization was emphasized. In light of this issue, the results obtained in the research demonstrate that, according to the *Smile Company* plan, all these workers who have to delegate their work, being made aware through training and practice of how they should delegate in order to always meet the established objectives, before even deciding who to delegate what to, must be well aware of the capabilities that this worker has to meet deadlines and ensure quality in the order.

In a third question, the focus was on an analysis of the competency to generate creativity, innovation and quality (*identifies, establishes priorities, and resolves individual or group problems*; *makes decisions in a timely manner based on a careful analysis of their effects*; and *recognizes the practical and ethical implications of the problem, recognizing those others affected by their decision*). Referring now to Section 7.2.2.2 , it was mentioned that SMEs are companies with great flexibility in adapting to the job market. They accept change and innovation more easily because they have the need to constantly adapt. At *Smile Company*, it is clear that innovation is always present, demonstrating the organization's capabilities and versatility to be a leader in the textile sector, developing partnerships and introducing innovative solutions, given that there is special attention at this level by certain departments so that the entire work cycle functions well to reach the final product.

At a fourth level, the study focused on the competency of results orientation (*commits to demanding and realistic objectives, being persistent in achieving the defined goals*; *shows interest in knowing the final work to which his/her tasks contribute*; and *manages his/her working time appropriately, showing concern for meeting the deadlines of the defined projects*), where unanimous identification was obtained from all respondents. This is because it is not relevant which department the person belongs to, since it is imperative that everyone knows how to meet deadlines, to ensure success in the entire order process, and to achieve the company's objectives together. This conclusion is automatically linked to Section 7.2.1.1, which refers to the fact that competencies consist of personal capabilities that allow people to perform their functions successfully, achieving results or completing tasks efficiently and effectively.

Subsequently, in a fifth question, the communication skills were analyzed (*ability to correctly use the Portuguese language to communicate effectively, orally, and in writing*; and *ability to correctly use the foreign language to adapt information and transmit knowledge*). It was noted that this skill also depends on each department and the need to use communication in the day-to-day activities of the role. There are departments that demonstrate total identification with these skills, but others that do not, which is normal as each function has its specificities. However, communication is necessary for all employees in order to avoid misunderstandings, so it could be further developed, thus avoiding possible conflicts due to misperception. Nowadays, it is increasingly common for individuals to have writing skills and language know-how. However, since the target group under study is a textile SME, the know-how in the identified departments stands out over this type of skills.

At a related point, the capacity to acquire and develop new skills was analyzed (*capacity and motivation to acquire and develop new skills proactively*), which proves what has been confirmed up to this point, as for this question too, there was a majority of workers identifying with the skills under study. And as mentioned, it is related to career management. The *Smile Company* gives its employees equal opportunities to evolve. It is up to each person to decide whether they really want to do it, or whether they just want to continue performing their function. As most people are open to change, workers are receptive and encouraged to acquire and develop new skills. This link to career management is in line with Section 7.2.3, where this same topic was addressed, talking about how career management can arise from performance evaluations where it is found that someone has skills that are not being well used and that would fit better in another type of role. This would make it possible to adjust each person's profile to a more suitable function.

Finally, the study focused on computer skills (*shows ability to use a computer to process, save, and manage information*; *shows solid knowledge of general Microsoft Office and easy handling of interactive programs essential for the creative development of the department's projects*; *shows ability to use a web browser to search and collect information in hypermedia format available on the internet, as well as use various services available on the internet such as email to receive and send messages and information*; and *shows ability to use a presentation program to present ideas and projects in order to involve groups of people and enable knowledge sharing*). All employees who need to handle computer equipment, even in different ways, have received training to develop the necessary skills. The results obtained demonstrate this reality, as everyone who identified with the parameters is in some way subject to technological equipment for their work. These types of skills start out specific, but become transversal. In what way? These skills, once acquired with the respective training, become transversal skills, which leads to a link to Section 7.2.1.1. As Ceitil [5] mentioned, the concept of transversal skills is distinguished from specific skills (associated with a specific function, profession, job), due to their transversality (non-contextualization) and transferability (acquired in one context and capable of being exercised in different contexts).

The results obtained were satisfactory. The *Smile Company*'s skills plan, in the eyes of its employees, is relatively well-designed, at least with regard to personal skills, as each employee has the knowledge necessary for their role, which they acquired through lectures and training provided. It is noticeable that there is a certain concern with the well-being of employees in their adaptation to their role, but there is also a correspondence of satisfaction and motivation of workers, with the conditions to which they are subjected in the exercise of their functions, as they end up feeling competent, which is a fundamental part of the company's success.

7.4 Final remarks

The research carried out contributed to understanding how important it is to have a good and elaborate planning process preceding competency management so that the entire process is successful in the end. It was possible to understand that this process brings countless advantages both in the production and development of each individual's work, as well as in the construction of a good culture and a good organizational climate, as well as helping in the formation and structuring of good work teams that are successful among themselves, since each individual will be in the role that they are truly suited to, with the training they need to achieve the necessary skills.

It is not always easy to make all employees in a company feel fulfilled because each person is different, and not everyone accepts having to participate in training outside of working hours, which, despite being for the development of their own skills, can be interpreted as abuse. However, this negative side does not apply to the company under study. What happens is the opposite, as employees after a training action feel even more capable and motivated to perform their role and help achieve the company's success, so that orders are delivered with quality and on time.

Each person has different types of abilities that are sometimes not as obvious as one might expect. But with proper development and monitoring, it is possible to reach the peak of these skills, so that later on the best performance of their role will be noticeable. It is essential that all employees are performing their duties to the best of their ability as there are employees who bring success to the company. And skills management is something that will make the company achieve its differentiation in the market where it operates, because the fact of retaining talent and having people with certain well-developed skills is its "weapon" against the competition, because technology and many other material resources are easy to be copied and adapted by other companies, while human capital, that is, people, are unique. This is why investing in the skills of each employee is not a cost, but an investment. It's what makes a company unique from its competition.

With this study, it is clear that competency management has many strengths and impact on an SME. As a small company, its competitive strength comes from its human capital as well as the ability to develop and retain talent within the company. Concern for employee well-being must be constant, because if they are not motivated and do not feel capable of performing their role to the best of their ability due to a lack of skills, they will not give their all for the company and it will be difficult for the company to reach its peak. And it is by using competency management that all these aspects can be assessed in order to achieve true success, preventing people from feeling that they are stagnating in their careers because the company is small, and from looking for large companies to achieve self-fulfillment.

References

[1] Brandão HP, Bahry CP (2014). Gestão por competências: Métodos e técnicas para mapeamento de competências. Revista Do Serviço Público 56(2): 179–194.

[2] Mlaouhi K, Gzara L, Cholez C (2021). Use of Competency Management Methods and Tools in Project-based Organizations. In: CIGI-Qualita21: Conférence Internationale Génie Industriel QUALITA, 2021, Grenoble, France, hal-03842413.

[3] Rosso F, García-Salirrosas EE (2023). Strategic trends in management by competencies: A bibliometric review. International Journal of Professional Business Review 8(9): 1–26,e03371.

[4] EU-Conexus (2024). EU-Conexus – Skills Survey Report. MS3, T3.1 Development of Signature Joint Masters on Smucs. Work Package Three (WP3) Joint Smucs Master Programme Development and Micro-credential Development, South East Technological University (SETU), European Union.

[5] Ceitil M (2010). Gestão E Desenvolvimento de Competências Lisboa, Sílabo.

[6] McClelland DC (1973). Testing for competence rather than for "Intelligence". American Psychologist 28(1): 1–14. https://doi.org/10.1037/h0034092.

[7] Brito L,M (2008). Gestão de Competências, Gestão do Conhecimento e Organizações de Aprendizagem: Instrumentos de apropriação pelo capital do saber do trabalhador. Cadernos de Educação 31 July-Dec. 203–225.
[8] Gasior M, Skowron L, Sak- Skowron M (2021). The importance of employees' competencies: A comparison between educational and business perspective. European Research Studies Journal XXIV (Special Issue 2): 681–694. https://doi.org/10.35808/ersj/2303.
[9] Groenewald ES, et al. (2024). Enhancing organizational performance through competency-based human resource management: A novel approach to performance evaluation. Educational Administration: Theory and Practice 30(8): 284–290. https://doi.org/10.53555/kuey.v30i8.7250.
[10] Tripathi K, Agrawal M (2014). Competency based management in organizational context: A literature review. Global Journal of Finance and Management 6(4): 349–356.
[11] Jardim J (2008). Programa de Desenvolvimento de Competências Pessoais E Sociais – Estudo Para a Promoção Do Sucesso Académico Lisboa: Instituto Piaget.
[12] Cassol A, Cintra R, Ruas R, Bassani D (2016). Evidenciação das Competências Gerenciais e a Influência na Atuação Estratégica das Micro e Pequenas Empresas de Santa Catarina. Revista da Micro E Pequena Empresa 10(3): 49–68. http://dx.doi.org/10.21714/19-82-25372016v10n3p4968.
[13] Kok D, et al. (2011). Do SMEs Create More and Better Jobs? EIM Business & Policy Research, European Communities, Competitiveness and Innovation Programme, Brussels: European Communities, Competitiveness and Innovation Programme.
[14] Melo PN, Machado CF (2020). O Livro de Ouro da Gestão de Recursos Humanos Nas Pequenas E Médias Empresas Lisboa: Editora RH.
[15] Melo P, Machado C (2015). Gestão de Recursos Humanos Nas Pequenas E Médias Empresas: Contextos, Métodos E Aplicações Lisboa: Editora RH.
[16] Roffia P, Moracchiato S (2021). Operationally defining family SMEs: A critical review. Journal of Small Business and Enterprise Development 28(2): 229–260. https://doi.org/10.1108/JSBED-11-2020-0399.
[17] Mandl I (2008). *Overview of Family Business Relevant Issues* – Final Report, Contract N° 30-CE-0164021/00–51 Vienna: Austrian Institute for SME Research and Turku School of Economics. European Commission, Enterprise and Industry Directorate-General.
[18] Handley K, Molloy C (2022). SME corporate governance: A literature review of informal mechanisms for governance. Meditari Accountancy Research 30(7): 310–333. https://doi.org/10.1108/MEDAR-06-2021-1321.
[19] Jones NB (2015). Impact of social media on small businesses. Journal of Small Business and Enterprise Development 22(4): 611–632.
[20] Chiavenato I (1990). Recursos Humanos (Edição Compacta) São Paulo: Editora Atlas S.A.
[21] Nakamori Y (2021). Converting Information into Knowledge. In: Knowledge Technology, Converting Data and Information into Values Ideas, Nakamori Y (ed) Singapore: Springer Briefs in Business. Springer, 57–73. https://doi.org/10.1007/978-981-16-3253-2_5.
[22] Quinn R, Thompson M, Faerman R, Mcgrath M (2003). Competências Gerenciais: Princípios E Aplicações Rio de Janeiro: Elsevier.
[23] Marcovitch J Ed. (1996). Tecnologia da Informação E Estratégia Empresarial Futura: São Paulo.
[24] Huda M, Sunardi M, Sumarsono TG (2022). The effect of remuneration, employee competency and leadership style on Civil Service performance. European Journal of Business and Management Research 7(5): 19–23. https://doi.org/10.24018/ejbmr.2022.7.5.409.
[25] Lana M,S, Ferreira VCP (2007). Gestão por competências: Impactos na gestão de pessoas. *Revista Estação Científica Online*, 3(Jan-Jun). https://estacio.periodicoscientificos.com.br/index.php/estacao cientifica/article/view/2570
[26] Pereira A (2013). Gestão de Competências – Desenvolvimento E Institucionalização de Um Portefólio de Competências-chave Numa Organização Estudantil Do Ensino Superior. Relatório de Estágio Do Mestrado Em Economia E Gestão de Recursos Humanos, Faculdade de Economia, Porto: da Universidade do.

Emin Taner Elmas

8 Prospective characteristics of contemporary engineers (using the approach of mechanical engineering)

Contribution and role of the mechanical engineer to organizational management and productivity

Abstract: If it becomes possible to train and raise the awareness of engineers, and if they are qualified and well educated in accordance with the statements mentioned in this chapter, it can be possible to achieve technical innovations and advances in many areas. In addition, it may be possible to make progress in the efficiency and productivity of industrial companies and factories. In this contribution, the productivity and organizational management will be concluded by a successful case study in industrial development arising from scientific studies and, finally, followed up by economic growth. Therefore, thanks to these studies, the welfare level of a given society can increase.

Those working in engineering mainly operate and manage domestic and global natural resources by considering national or international interests and contribute to increase the productivity of industrial applications.

All engineers, regardless of their specific profession or branch, must be conscious of society and the environment while carrying out their engineering tasks and behave responsibly.

This chapter consists of lecture notes that I have used in recent years in my "Introduction to Mechanical Engineering" course while I was a lecturer at the university. This chapter explains scientifically the kind of fundamental characteristics and educational background level an engineer should have with respect to business, industrial, and academic life.

I have tried to describe the science of mechanical engineering and its research interests and scope of business as well as the duties engineers undertake during their professional life. I have explained the procedure to follow to be a well-educated, skilled engineer and to have a solid engineering career. I also provide some important factors for engineers to be successful in their business life based on the knowledge and experience I have accumulated over the course of my academic and industrial professional life. I have tried to emphasize the prospective fundamental characteristics of a newly minted engineer in today's world. While presenting that information, I also consider the main objectives and requirements of learning outcomes for accreditation boards for engineering education.

In spite of the fact that this chapter was written with mechanical engineering in mind, the content and descriptions are comprehensive and valid for most other engi-

https://doi.org/10.1515/9783112206775-008

neering branches. I believe that this chapter will be very helpful for all students of engineering in universities and for graduate engineers working in all industrial or academic fields. It will also enlighten and point the way for high school students who intend to enter engineering faculties and to become engineers.

This chapter was prepared and written in presentation format, so it is easy to read and follow (based on the work of Elmas [1]). I hope all readers will find it useful.

8.1 The objective of mechanical engineering

Engineering is an occupation whose goal is to develop methods for the economic use of resources, potentials, and materials in nature, for the benefit and welfare of human beings. This is accomplished by a reasoning process of experience and practical knowledge gained from the fundamental sciences and mathematics.

The main objective of engineering is to develop the existing technology and bring into existence new systems for human use by introducing the design and production of innovative technical applications in consideration of scientific knowledge.

The most important infrastructure in terms of producing unique technology is fundamental scientific research and qualified manpower and brain power.

The capabilities and contributions of engineers are so important that those capacities affect directly the assessment by educated people of the potential of a society or country and even the entire world.

During the engineering execution period, engineers generally think, search, plan, design, and conceive projects and implement, produce, and manage the related industrial and business activities. The research capabilities of engineers are as important as their scientific knowledge and field experience.

In addition to producing unique technology, engineers should carry out tasks related to technology transfer, adaptation, matching, customization, implementation, and usage.

The place of mechanical engineering in society and its importance can only be described by introducing and presenting in detail the jobs involved in the field. A strong relationship exists between a country's reconstruction and industrialization, and this relationship cannot be ignored. Industrial progress is one of the cornerstones of community development. Mechanical engineering is one of the most important driving forces for periods of industrialization and development for a country and for the world.

Mechanical engineering as an occupation establishes a fundamental bridge for each applicable area of contemporary science and technology, and engineers perform the following duties.

People working in mechanical engineering mainly use and manage domestic and global natural resources by considering the national or international interests; they also contribute to increasing the productivity of industry. They generally carry out technical studies and scientific research to facilitate the industrial development of their own country in parallel to national requirements. Other related activities include engineering services, for example, searching activities, inspections, audits, preparing and implementing projects, and writing reports.

All of these activities make it possible to offer services in accordance with the related technical regulations, norms, and standards and allow for their supervision, surveillance, checking, approval, and review. Expertise and review studies help resolve technical and financial issues that arise from various industrial applications.

8.2 Work scope and duties of mechanical engineers

Mechanical engineers are engaged in a very broad range of activities in a variety of fields, which are described in detail in what follows. After graduation, newly minted engineers can be employed in many different industrial and academic areas. As can be clearly understood from the following list, mechanical engineers work in almost all branches of industry, so they hold great importance, playing a vital role and making key contributions to organizational management and productivity. Most industrial companies, technology institutions, governmental bodies, universities, production plants, and factories employ mechanical engineers for various kinds of duties as follows:

- Design, manufacturing, installation, and commissioning of hot water boilers, superheated water boilers, saturated and superheated steam boilers, waste heat recovery boilers, and power boilers
- Plant engineering and sanitary engineering
- Power piping
- Automotive industry
- Aerospace and aviation industries
- Aeronautic applications
- Pressure vessel design, manufacture, and assembly
- Storage tank design, manufacture, and on-site assembly
- Electronics
- Defense industry
- Heating, ventilation, and air conditioning (HVAC)
- Manufacture of spare parts for various industries
- Metal machining works
- Manufacturing by welding seams
- Sheet metal forming and pressing
- Casting works and metallurgical applications
- Molding

- Medical technology and applications
- Production of medical devices and related machines and equipment
- Renewable energy sources and applications (e.g., solar, wind, geothermal, biomass, and hydraulic)
- Academia
- Research in research and development (R&D) companies and associations
- Electrical energy production and distribution systems
- Thermal power plants
- Hydraulic power plants
- Nuclear power plants
- Application of environmental technologies
- Solid waste treatment plants
- Domestic wastewater treatment plants
- Chemical wastewater treatment plants
- Pure water treatment plants for preparing feed water
- Recovery techniques
- Recycling
- Iron and steel industry
- Mining
- Manufacture of general-purpose machines
- Domestic appliances
- Manufacture of lifting machines
- Machines for transport technology
- Turbomachinery
- Internal combustion engines
- Robotic applications
- Cement, ceramic, and glass industries
- Civil constructional commitment works
- Food and beverage industry
- Pulp and paper industry
- Paint industry
- Agricultural machinery
- Chemical industry
- Evaporation, distillation, desalination, and extraction plants
- Coating and plating industry
- Petrochemical technology
- Refineries
- Rocket and jet-propulsion systems
- Nanotechnology
- Work in foreign countries (assuming the requisite experience and knowledge of local language)

These fields of work may be different for mechanical engineers depending on the production techniques used.

Example:
- Mass production (flow-shop manufacturing)
- Job-order production (manufacturing on the order base – the most common applications are contracting jobs)

Such jobs can only be related to manufacturing orders or complete contracting orders, including the technical and commercial activities consisting of the design, manufacturing, installation, and commissioning stages.

Mechanical engineers generally work in the following departments depending on the specific features of field:
- Thermal design
- Mechanical design
- Construction design
- Projects and drawings
- Planning
- Manufacturing and production
- Assembly, construction, and installation
- Field and work site applications
- Factory/workshop applications
- Continuous operation of industrial plants
- Sales and marketing
- Quotation/bidding on price, delivery time, and conditions for payment and delivery, preparing and working out costs
- Claims, remuneration, and payment schedule
- Maintenance and repair activities
- Project management
- Production planning
- R&D
- Product development
- Quality control
- Quality assurance
- Unit testing
- Management and organization
- Purchasing
- Business development
- Supplier management
- Warehouse management
- Logistics
- Supply-chain management
- Computer and software applications

- Technical education
- Occupational health and safety

Mechanical engineers commonly hold the following job titles in the aforementioned areas, depending on the characteristics of the given area and enterprise structure:

- Engineer (i.e., project engineer)
- Junior engineer
- Senior engineer
- Chief engineer
- Supervisor (i.e., manufacturing supervisor)
- Department chief
- Project manager
- Sales manager
- Manufacturing /production manager
- Design manager
- Testing manager
- Quality manager
- Contracting manager
- Assistant manager or deputy department manager
- Department manager
- Administrator
- Deputy general manager
- General manager
- Managing director
- Coordinator
- Auditor
- Lead auditor
- Member of board of directors
- Executive member of board of directors
- Chairman of executive board
- Executive committee member
- Chief executive officer (CEO)
- Research assistant (in research associations, institutes, and academia)
- Teaching assistant
- Research associate
- Researcher or research participant
- Academician (university lecturer, instructor, reader, faculty member, assistant professor, associate professor, adjunct professor, professor)
- Member of editorial board of scientific journals
- Editor of scientific journal or publication
- Peer reviewer, reviewer
- Observer

- Project referee
- Expert

8.3 How to be a well-educated engineer and to have a solid engineering career

An engineer should have the following qualifications in ensuring support and a solid career.

8.3.1 University degree

Graduates with a bachelor's degree (B.Sc.) have a high-grade point average (3.00–4.00 or a minimum of 75/100). This level is widely recognized internationally as a mark of excellence in one's studies.

8.3.2 Foreign language

Graduates need to have good knowledge of a foreign language, in particular, an advanced level of knowledge of English. Technical English at a very advanced level is a must in addition to an ability to write academic reports. Knowledge of business English is also very important.

8.3.3 Computer knowledge

- Especially current technical in one's field of interest as an engineer
- Microsoft Office: Windows, Word, Excel, PowerPoint, and other programs
- Advance Internet search capabilities using relevant technical and scientific keywords
- Computer-aided design and computer-aided manufacturing
- Finite-element methods using ANSYS and similar software, finite-difference methods, finite-volume methods by computational fluid dynamics analysis using FLUENT and similar software
- COMSOL, ASPEN, and similar software packages, CAESAR, PV ELITE, COMPRESS, AUTO PIPE, GT PRO, and others
- Design software such as AUTOCAD, SOLIDWORKS, AutoPLANT, CATIA, and related software

8.3.4 Education following graduation

8.3.4.1 Graduate school

Engineers holding just a bachelor of science (B.Sc.) degree should look to advance their theoretical scientific knowledge in some sort of graduate programs in the branch of science most closely related to their major department. One can pursue a master of science (M.Sc.) or a doctor of philosophy (Ph.D.) degree; alternatively, an engineer might wish to pursue a master's in business administration (MBA).

In addition to graduate education, engineers should also consider a certification program for personal and professional development. In addition, intensive supplementary occupational training courses and job training programs must also be taken to remain current in one's professional knowledge and business capabilities.

8.4 Important factors for success in engineering

Engineers should have the following individual personality traits to be successful in their careers:
- Be able to take on responsibilities
- Be able to take initiative
- Be open to self-improvement
- Be able to make quick and correct decisions
- Be hardworking
- Be practical (be able to offer practical solutions to problems)
- Be patient
- Be open to dialogue and have good communication skills
- Be enterprising and like being challenged
- Have general cultural knowledge at a high level for both social and technical areas
- Apply oneself in one's studies and on the job
- Be respectful of experienced colleagues
- Remain current on innovations and technical developments
- Behave rationally and with awareness
- Be broad-minded
- Be results-oriented
- Be an excellent observer

To be a good engineer, students in engineering programs must understand that all lectures and classes are of equal importance. There are important and serious reasons for every class lecture. For this reason, all lessons must be learned at the highest level and classes must be attended regularly.

8.5 Prospective characteristics of a newly minted engineer

The following 14 items are fully compatible with the general requirements of accreditation boards for engineering education.

8.5.1 Characteristic 1

To be able to establish connections between mathematics, fundamental sciences, engineering sciences, and engineering applications:

An engineer should have the following capabilities to solve problems that may arise over the course of engineering applications:

- To know the fundamental scientific concepts, laws, notions, terms, and basic principles
- To be able to specify the technical parameters that identify the operating system and to be able to formulate mathematical models
- To be able to make engineering assumptions and employ related approaches that are required for simplification of the theoretical models

8.5.2 Characteristic 2

To be able to approach engineering problems and requirements systematically through the use of scientific principles and to apply those principles to arrive at a solution:

- To be able to define an issue that requires engineering services, with the help of the engineering context and technical parameters
- To be able to determine the input-output parameters of problems by identifying the cause-and-effect relationship
- To be able to break down an engineering problem into subproblems that can be simplified and solved independently one by one, and then to be able to combine the results to obtain one solution
- To produce practically applicable solutions

8.5.3 Characteristic 3

To have the ability to attain the required technical knowledge and identify its source and to be able to comment on them:

An engineer should have the following qualifications, which relate to the examination and analysis of previous studies and experiments, in order to have advanced knowledge on subjects, which are carried out to find solutions to engineering problems:
- To be able to identify keywords regarding a particular area of interest
- To have the ability to access the required technical and scientific reference works
- To have the ability to identify useful data as determined from sources
- To gain the ability to edit data and knowledge obtained from reference works (identifying, classifying, rearranging, and reporting on them)

8.5.4 Characteristic 4

To be able to carry out the engineering design:
- To be able to determine the constituents of the design period and to be aware of the design systematic
- To be able to carry out conceptual/predesign studies
- To be able to specify the criteria and parameters of a design problem as well as to designate, assign, and make use of them
- To gain experience doing detailed design calculations
- To gain an interdisciplinary perspective, point of view, and logic
- To be able to use methods that can access optimum solutions in design works
- To be able to consider the national, international, and local technical norms and standards affecting design activities
- To be able to explain an applied design work by stating its strengths and weaknesses
- To be able to determine design items, for example, performance, capacity, working life, durability, system reliability, capital cost, operating cost, feasibility, productivity, efficiency, effectiveness, consumption, and payback period
- To have enough knowledge of numerical solution methods in engineering

8.5.5 Characteristic 5

To be able to follow the publications, literature, and technology related to one's areas of interest and associated subjects:
- To be aware of the relevant scientific and technological publications, chapters, books, articles, papers, presentations, and journals and to know how to track them down

- To be informed about the scientific and occupational meetings related to the different sections of mechanical engineering or other related engineering branches
- To be informed about specialized occupational exhibitions related to the different sections of mechanical engineering and other engineering branches
- To be informed about national and international technical norms and standards relevant to production and control activities

8.5.6 Characteristic 6

To be a lifelong learner:

When developments in engineering, science, and technology are considered, during one's career:
- To be motivated to update one's professional qualifications
- To challenge and examine continuously the knowledge acquired during the university graduation
- To be current on the latest developments and innovations in technical applications, as well as to reevaluate and update one's knowledge

8.5.7 Characteristic 7

To have effective written and verbal communication skills:
- To be able to prepare documents, certificates, reports, and presentations related to the engineering services carried out by oneself
- To be able to communicate in writing and orally while carrying out one's engineering tasks
- To have advanced knowledge of the technical terms and technological vocabulary in one's field, both in one's native language and in English, which are integral and inseparable parts of mechanical engineering

8.5.8 Characteristic 8

To be able to find unique solutions to specific problems:
- An engineer should be able to find creative solutions using various kinds of engineering systems, technological applications, and components.
- When required, an engineer should be able to find extraordinary solutions and state the strengths and weaknesses of those solutions.
- An engineer should be able to produce an array of solution options and define their strengths and weaknesses.

8.5.9 Characteristic 9

To be aware of the social, environmental, and economic effects while performing occupational tasks:
- To be aware of the regulations regarding occupational health and safety, environmental influences, and economic parameters governing engineering activities
- To be sensitive to and conscious of society, the environment, and nature while carrying out engineering tasks, as well as to behave responsibly toward those subjects
- To protect and contribute to one's occupation and occupational organization, such as engineering associations, in order to advance the field and to enhance the reputation of engineering as a profession

8.5.10 Characteristic 10

To be open to new ideas:
- Engineers should possess the kind of scientific education that will allow them to follow and implement the latest technological developments in the area of mechanical engineering
- To be open to the latest technological methodologies and opinions, and to be willing to adopt them
- To be able to use the latest methods, instruments, and techniques in engineering
- To have a broad general knowledge of production methods

8.5.11 Characteristic 11

To be able to work within a team setting (disciplinary, interdisciplinary, and multidisciplinary teamwork abilities):

An engineer should have the following abilities to carry out one's duties as required by engineering practices:
- To be able to provide constructive support for one's team
- To be able to assign disciplinary and interdisciplinary tasks, to perform them seriously, and to fulfill one's responsibilities in a timely manner
- To be able to share one's knowledge with team members, to consider other solution suggestions, to remain open to others' ideas, and to be able to discuss subjects on the basis of engineering so as to carry out a project in collaboration with one's team members
- To arrive at a collaborative solution considering other aspects of a project or by taking into account other scientific and technical disciplines

- To have a solid understanding of the fundamental principles of project management, including preparation, planning, timing, adoption, execution, finalization, and feedback

8.5.12 Characteristic 12

To have a strong level of comprehension of engineering ethics:
- To be aware of the existence of engineering ethics, to know about basic ethics, and to live up to basic ethical standards
- To possess general ethical values
- To respect the occupational hierarchy
- To respect the superintendent

8.5.13 Characteristic 13

To have the ability to carry out tests and experiments:
- To be able to conduct the experimental practices of mechanical engineering and to be able to carry out the required engineering applications in accordance with those practices
- To know the appropriate conceptions of accuracy, reliability, and calibration of measurement methods and devices, in addition to the conceptual fields of inaccuracy, tolerance, sensibility, and precision with respect to given measured dimensions and technical and physical parameters
- To be able to design the experiments and testing apparatus properly
- To possess the requisite knowledge of measurement and tracking techniques
- To be able to comment on experimental results in light of fundamental scientific principles, theories, and laws and to translate them into practical applications

8.5.14 Characteristic 14

To possess the requisite qualifications for being a pioneer and leader for the professional career opportunities and occupational subjects:
- To be able to motivate team members so as to provide services and use resources most efficiently in mechanical engineering issues
- To be able to take the initiative on the strength of one's engineering knowledge and experience; an engineer should be able to make decisions in situations where crucial information is uncertain, unclear, indefinite, vague, lacking, imperfect, weak, inadequate, unsatisfactory, or inconclusive
- To be able to produce novel and reasonable solutions and to be able to present them to others

- To be able to work creatively and productively in the context of working conditions with a variety of people

The preceding 14 items describe the traits of a modern, well-educated engineer with a solid career. Those items can be summarized as follows:

1. To be able to establish connections between mathematics, fundamental sciences, engineering sciences, and engineering applications
2. To be able to approach engineering problems and requirements using scientific principles and arrive at a solution
3. To be able to access required technical knowledge and its source and be able to comment on them
4. To be able to realize engineering design
5. To be able to follow the publications, literature, and technology related to one's areas of interest
6. To be a lifelong learner
7. To be able to communicate effectively in writing and orally
8. To be able to find unique solutions to specific problems
9. To be aware of social, environmental, and economic effects of one's job
10. To be open to new ideas
11. To be able to work within a team setting (disciplinary, interdisciplinary, and multidisciplinary teamwork abilities)
12. To have a deep level of comprehension of engineering ethics
13. To be able to conduct tests and experiments
14. To be a pioneer and leader with respect to career opportunities and occupational areas of interest

8.6 Contribution and role of mechanical engineers to organizational management and productivity

Engineers who are trainable and educated in accordance with the statements given in this chapter will be successful, and their activities will lead to industrial development and economic growth. In turn, the welfare level of the society in which such engineers work will rise.

It is also possible to achieve technical innovations and advances in many areas. In addition, it may be possible to make gains in efficiency and in the productivity of industrial companies and factories.

If companies, corporations, institutions, and universities are organized in line with the concepts outlined in this chapter, they can attain an institutional structure. Mechanical and other engineers in various engineering professions and branches will have an important and serious role to play in the technical and administrative management of companies, corporations, factories, institutes, and universities. Engineers will make significant contributions to organizational management and, therefore, to productivity through the application of their skills and knowledge gained through professional experience since mechanical engineering provides them with an analytical view and exceptional problem-solving skills.

In addition to the aforementioned issues, mechanical engineers can work in almost all departments in an organization. In addition, they can work at each level of an organization and advance in their careers. These characteristics of the engineering profession allow for other contributions to organizational productivity.

All organizations need engineers from varied backgrounds with miscellaneous skills. Some types of engineers have deep, theoretical, scientific knowledge in many technical subjects in general, so they can be pioneers and leaders in many occupational fields and manage organizational structures. The second type of engineers focuses on a few subjects in detail, so engineers are occupational experts in their fields of interest. The remaining type, the third type, of engineers has direct practical experience in some application areas, such as in factory workshops and on-job sites. All the different types of engineers have important contributions to make to the productivity of organizations.

I would like to explain one particular provision by giving an unusual and impressive example to highlight the importance of engineers for all kinds of industrial, occupational, investigational, institutional, and academic areas, including medical applications, as described subsequently in Section 8.6.1.

8.6.1 A specific example of a medical technique in engineering and science

By giving a specific example in this section, it is possible to explain impressively and clearly the contribution and role of mechanical engineers, as well as engineers in other related engineering branches, to organizational management and productivity by emphasizing *the level of their participation*, even for a different and important science like medicine, and therefore for the importance of health of human beings and also for the sustainability of life. Moreover, there is a strong connection between the applications of medical techniques and the development of society and economic welfare.

As an applied branch of engineering and science, medical technology has a broad range of uses. Medical devices, machines, equipment, and apparatus are widely used in medical diagnosis and treatment. Therefore, medical technology can be seen as an inseparable part of medical science.

All over the world and also at the various space locations of the universe, changing from the robotic surgery to the space medicine, there are many fields of application of medical technology. This is an interdisciplinary technology and science, bringing together researchers, academics, engineers, and scientists from different fields of expertise, that is, mainly mechanical engineering, electrical-electronics engineering, computer engineering, chemical engineering, engineering physics, astronomy, fundamental physics, fundamental chemistry, biology, mathematics, pharmacy, dentistry, and medicine itself.

This particular study (medical technique in engineering and science) mentioned in this section was written to explain the importance of engineering and the fundamental science branches for the implementation of medical science using medical technology. Many related topics can be introduced and itemized by separating them into main groups and their detailed contents in order to make clear the required subjects so that engineers and scientists can focus on their fields of interest in relation to the corresponding medical technology.

Hence, this section of the study is aimed at creating an awareness of medical technology and, moreover, to encourage engineers and scientists to study those subjects. The chapter also makes clear that engineering technology is of great importance in medical science and greatly affects human health.

By proposing various kinds of research in medical technology, researchers, academics, engineers, and scientists can establish connections between the subjects of medical technology and their expertise, so they can study various areas. Once they study medical technology, many engineers and scientists will start to invent or develop medical devices, machines, equipment, and apparatus that can provide medical diagnosis and treatment. Moreover, these kinds of studies that require a multidisciplinary approach can also ensure the local production of medical devices, machines, equipment, and apparatus in each country and contribute to the development of engineering and science in connection with economic growth by increasing total productivity.

In conclusion, and as can be clearly understood from the preceding example, it is possible to assert that engineers in various professions will make significant contributions to the management and productivity of many different kinds of organizations.

Reference

[1] Elmas ET (2014). Çağımızın Mühendisinden Beklenenler, Gece Kitaplığı.

J. Paulo Davim

9 Productivity and management: a bibliometric analysis

Abstract: Productivity is a fundamental factor for the growth and development of agriculture, industry, and services. The Scopus database was used for the bibliometric analysis, based on the terms {productivity} and {management}. The better results obtained show in the function of the number of documents produced: year – 2024; source – *Agricultural Water Management*; author – Lal R.; affiliation – Chinese Academy of Sciences; country – USA; document type – article; scientific area – *Agricultural and Biological Sciences*; and funding support – National Natural Science Foundation of China.

According to the *Bureau of Labor Statistics*, productivity is defined as "a measure of economic performance that compares the quantity of goods and services produced (output) with the quantity of inputs used to produce those goods and services." Today, managers, engineers, and other professionals ask a lot about which strategies would produce the greatest productivity gains [1–3].

The bibliometric analysis used the database Scopus/Elsevier to search for the documents. Using the terms {productivity} and {management}, TITLE-ABS-KEY ({productivity} and {management}), 117,105 documents were identified (a search carried out on May 26, 2025).

The results obtained in documents are listed in Table 9.1, for the first 10 positions concerning year, source, author, affiliation, country, document type, scientific area, and funding support. The better results obtained show in the function of the number of documents produced: year: 2024 (10,603), followed by the years 2023 (7,918) and 2022 (7,025); source: *Agricultural Water Management* – Elsevier (963) followed by *Sustainability Switzerland* – MDPI (939) and *Forest Ecology and Management* – Elsevier (877); authors: Lal R. – The Ohio State University (187), followed by Hossain A. – Bangladesh Wheat and Maize Research Institute (86) and Babu S. – ICAR – Indian Agricultural Research Institute (83); affiliation: Chinese Academy of Sciences (1955) followed by Commonwealth Scientific and Industrial Research Organisation (950) and Ministry of Education of China (11,191); country: USA (27,779) followed by India (12,130) and China (696); document type: article (78,229) followed by conference paper (19,783) and review (8,888); scientific area: *Agricultural and Biological Sciences* (36,153) followed by *Environmental Science* (28,011) and *Engineering* (25,932); and funding support: National Natural Science Foundation of China (3,928) followed by the Ministry of Science and Technology of China (3,057) and European Commission (2035).

https://doi.org/10.1515/9783112206775-009

Table 9.1: Documents (117,105) by item.

	Year	Source	Author	Affiliation
1	2024 (10,603)	*Agricultural Water Management* (963)	Lal, R. (187)	Chinese Academy of Sciences (1955)
2	2023 (7,918)	*Sustainability Switzerland* (939)	Hossain, A. (86)	Commonwealth Scientific and Industrial Research Organisation (950)
3	2022 (7,025)	*Forest Ecology and Management* (877)	Babu, S. (83)	Ministry of Education of China (882)
4	2021 (6,766)	*Science of the Total Environment* (825)	Jat, M.L. (80)	Indian Council of Agricultural Research (880)
5	2020 (5,967)	*IOP Conference Series: Earth and Environmental Science* (704)	Giller, K.E. (74)	USDA Agricultural Research Service (878)
6	2019 (5,563)	*PLOS One* (615)	Wani, S.P. (66)	Wageningen University & Research (861)
7	2018 (4,664)	*Acta Horticulturae* (563)	Choudhary, A.K. (63)	University of Chinese Academy of Sciences (786)
8	2017 (4,268)	*Journal of Environmental Management* (559)	Das, T.K. (63)	ICAR – Indian Agricultural Res. Institute, New Delhi (767)
9	2016 (4,035)	*Field Crops Research* (555)	Das, A. (60)	University of Florida (680)
10	2015 (3,759)	*Agriculture Ecosystems and Environment* (539)	Siddique, K.H.M. (60)	China Agricultural University (651)
	Country	**Type**	**Area**	**Funding support**
1	USA (27,779)	Article (78,229)	*Agricultural and Biological Sciences* (36,153)	National Natural Science Foundation of China (3,928)
2	India (12,130)	Conference paper (19,783)	*Environmental Science* (28,011)	Ministry of Science and Technology of China (3,057)
3	China (11,191)	Review (8,888)	*Engineering* (25,932)	European Commission (2035)
4	UK (6,762)	Book chapter (6,145)	Medicine (15,148)	National Key R&D Program of China (1349)
5	Australia (5,960)	Book (961)	Computer science (14,349)	National Science Foundation (1193)

Table 9.1 (continued)

	Year	Source	Author	Affiliation
6	Canada (4,841)	Note (861)	*Business, Management and Accounting* (13,727)	Conselho Nacional de Desenv. Científico e Tecnológico (973)
7	Brazil (4,432)	Short survey (722)	Social sciences (11,940)	U.S. Department of Agriculture (967)
8	Germany (4,157)	Conference review (555)	*Earth and Planetary Sciences* (9,410)	Coordenação de Aperf. de Pessoal de Nível Superior (870)
9	Spain (3,309)	Editorial (515)	*Energy* (6,606)	UK Research and Innovation (862)
10	Italy (3,197)	Letter (249)	*Decision Sciences* (5,485)	Horizon 2020 Framework Programme (676)

Source: Scopus/Elsevier, May 26, 2025.

References

[1] Perry E (2023). What is productivity? Definition and ways to improve, BetterUp. https://www.betterup.com/blog/what-is-productivity (accessed May 26th, 2025)

[2] Machado C, Davim JP ed. (2019). Management Science: Foundations and Innovations, Springer. https://link.springer.com/book/10.1007/978-3-030-13229-3.

[3] Machado C, Davim JP ed. (2017). Productivity and Organizational Management, DE Gruyter. https://www.degruyterbrill.com/document/doi/10.1515/9783110355796/html.

Index

360° evaluation 91

ability-motivation-opportunity 95
adaptation 116, 128, 154, 157, 163
affiliation 183
agriculture 183
anticipation 116, 130
assessment procedures 95
author 183

beggar-my-neighbor effects 83
boundary spanning 92
business value 2
business value generation 10

capabilities 151, 153–154, 158, 160–162
capital measurement system 3
career development 155
change 154–157, 161–162
classes 174
clinical expertise 123
communication 155, 158, 162
communication skills 177
compensation 156
competencies 151, 153, 161
competency development 114, 127
competency management 151–153, 156–157, 162–164
competitive market 151
competitiveness 78, 82
country 183
customization 116, 129

deliverology 89
develop 151–154, 156–157, 160–164
differentiation 130
disciplinary 178
doctors and nurses 120
doctors availability ratio 127
doctors' contractual ties 127
document type 183

economic efficiency 79, 88, 93, 96, 102
economic growth 180
economic productivity 78
educare 94
effectiveness 153
efficiency 49–51, 68–69, 180
efficient economy 88, 103
efficient market 103
efficient society 88, 103
emotional intelligence 154, 160
employees 151–152, 154–164
engineering 168
engineering design 176
engineering ethics 179
expected performance 127
experiment 179

family businesses 154
family involvement 155
feedback 153, 156
financial incentives 88
flexibility 154–155, 158, 161
flexibility-by-consent 102
flexible labor market 84
flexible labor markets 82
Fordist Weberian paradigm 81
front-back office design 89
functional skills 154
fundamental sciences 175
funding support 183

globalization dynamics 113

health field 117
health system 109
HHR assignment 120
HHR competency development 118–119
HHR distribution 118–119, 128
HHR labor 117
HHR management 120, 128
HHR management process 118
high performance 81
high-involvement work system 93
high-performance work system 86, 93
HRM 21–22, 26–27, 30, 44–46
HRM perspective 133
HRM policies and practices 112
human capital 2, 111, 136, 155–157, 164

https://doi.org/10.1515/9783112206775-010

human capital investment 1
human health 182
human resource management 109, 133–134

inducare 95
industrial development 180
industry 183
informality 155
innovation 155, 158, 161
innovation-by-agreement 84, 102–103
interdisciplinary 178
internal market 96
ISO 9001 22

job satisfaction 159
job-order production 171

key competencies 157
knowing how to be 154, 160
knowing how to behave 154, 160

labor markets 82
labor productivity 84
labor-intensive employment 100
leadership position 156, 161
learner 177
learning capacity 152
lectures 174
level of impact 5
lifetime employment 80–81, 83–84, 102
Lindbeck-Snower insider-outsider claim 83
literature 176
LMX theory 83, 86

macroeconomic austerity 82
management 151, 157, 183
management competency 137, 146
management of competencies 152
management structures 85
management tools 50, 66–67
manufacturing 171
market criteria 94
Martin Programme on the Impacts of Future Technology 99
mass production 171
mathematics 175
measurement of human capital 17
mechanical engineer 169
medical technique 181
medical turnover 126
methodological knowledge 153
mobilization 116, 129
model 7
multinational corporations 99

Nep Public Management paradigm 96

observed performance 127
occupied vacancies 122
open vacancies 122
operational managers 83
organizational change 86
organizational climate 163
organizational department 68
organizational goals 156
organizational knowledge 153
organizational logic 86, 92
organizational management 167, 169
organizational performance 112
overground economy 100

performance 21–22, 25–27, 30, 33–34, 36, 38, 45–46, 111
performance appraisal 90
performance assessment 91
performance evaluation 153, 157
performance evaluations 156, 162
performance practices 131
personal 153–154, 160–163
personnel management 114
person's ability 153
Portuguese NHS 110
Portuguese NHS supply 121
post-Fordist flexible production model 97
problem-solving skills 181
production organization 67
productivity 4, 33, 49, 51, 54, 58, 68–69, 135, 151, 167, 169, 183
productivity enhancement interventions 133, 137
productivity improvement 109
productivity ratio 135
professional career 179
profit sharing 81
profitability 4, 99
prospective characteristics 175
public expenditure 79

public health 124
public investment 78
public sector employees 90

QMS 22–29, 31, 33–35, 37–39, 42, 44–46
quality 21

recruitment 121, 153, 159
relational coordination 92
relational models 88
relationship with workers 130
results orientation 162
retain 152, 157, 164
return on investment in human capital 10
rewards management 153
ROI^{CH} 17

science 168
scientific area 183
scientific knowledge 181
sector 17
selection 153, 159
self-regulation procedure 119
service organization 67–68
services 183
sharing 129
SHRM 113
SHRM in the Portuguese NHS 117
skills 151–157, 159–164
skills management 151–153, 156–157, 164
SME 151–152, 154–155, 157, 162, 164
SMEs 151–152, 154–158, 161–162
social domains 87
social efficiency 77, 79–80, 93, 101
social justice 79
social skills 154, 161
solutions 177
source 183
specific skills 152, 163
Sterns' revolving door effect 97
strategic contribution 1
strategic control points 134
strategic dimension of HRM 110
strategic human resource management 110
structural equations 13
structural unemployment 77
success 174
sustainable competitive advantages 111
sustained productivity 81
systemic thinking 112

talent 151, 156–157, 164
team setting 178
team spirit 153
team work 146
technical innovations 167
technical knowledge 152–153
technological unemployment 77, 98–100
technology 168
total productivity 182
training 151, 153, 155–156, 159, 161, 163
training management 153
transactional leadership 85
transferability 152, 163
transversal skills 152, 163
transversality 152, 163

unreconstructed Weberian and Fordist paradigm 97

value creation 4
versatility 154, 158, 162

work performance 91
worker 121
worker commitment and autonomy 112
worker distribution 114
work-life balance 93, 102–103
workplace 152

year 183

www.ingramcontent.com/pod-product-compliance
Lightning Source LLC
LaVergne TN
LVHW080653011026
849160LV00014B/1412
9783119147323